Methods in the Theory of Hereditarily Indecomposable Banach Spaces

MEMOIRS
of the
American Mathematical Society

Number 806

Methods in the Theory of Hereditarily Indecomposable Banach Spaces

Spiros A. Argyros
Andreas Tolias

July 2004 • Volume 170 • Number 806 (third of 4 numbers) • ISSN 0065-9266

American Mathematical Society
Providence, Rhode Island

2000 *Mathematics Subject Classification.*
Primary 46B20, 46B26, 46B03.

Library of Congress Cataloging-in-Publication Data

Argyros, S. (Spiros), 1950–
Methods in the theory of hereditarily indecomposable Banach spaces / Spiros, A. Argyros, Andreas Tolias.
p. cm. — (Memoirs of the American Mathematical Society, ISSN 0065-9266 ; no. 806)
"July 2004, Volume 170, Number 806 (third of 4 numbers)."
Includes bibliographical references.
ISBN 0-8218-3521-1 (alk. paper)
1. Normed linear spaces. 2. Banach spaces. I. Tolias, Andreas, 1974– II. Title. III. Series.
QA3.A57 no. 806
[QA322.2]
510 s–dc22
[515′.732] 2004046103

Memoirs of the American Mathematical Society

This journal is devoted entirely to research in pure and applied mathematics.

Subscription information. The 2004 subscription begins with volume 167 and consists of six mailings, each containing one or more numbers. Subscription prices for 2004 are $583 list, $466 institutional member. A late charge of 10% of the subscription price will be imposed on orders received from nonmembers after January 1 of the subscription year. Subscribers outside the United States and India must pay a postage surcharge of $31; subscribers in India must pay a postage surcharge of $43. Expedited delivery to destinations in North America $35; elsewhere $130. Each number may be ordered separately; *please specify number* when ordering an individual number. For prices and titles of recently released numbers, see the New Publications sections of the *Notices of the American Mathematical Society.*

Back number information. For back issues see the *AMS Catalog of Publications.*

Subscriptions and orders should be addressed to the American Mathematical Society, P. O. Box 845904, Boston, MA 02284-5904, USA. *All orders must be accompanied by payment.* Other correspondence should be addressed to 201 Charles Street, Providence, RI 02904-2294, USA.

Memoirs of the American Mathematical Society is published bimonthly (each volume consisting usually of more than one number) by the American Mathematical Society at 201 Charles Street, Providence, RI 02904-2294, USA. Periodicals postage paid at Providence, RI. Postmaster: Send address changes to Memoirs, American Mathematical Society, 201 Charles Street, Providence, RI 02904-2294, USA.

This publication is indexed in *Science Citation Index®*, *SciSearch®*, *Research Alert®*, *CompuMath Citation Index®*, *Current Contents®/Physical, Chemical & Earth Sciences.*
Printed in the United States of America.

∞ The paper used in this book is acid-free and falls within the guidelines established to ensure permanence and durability.
Visit the AMS home page at `http://www.ams.org/`

10 9 8 7 6 5 4 3 2 1 09 08 07 06 05 04

Contents

ABSTRACT. A general method producing Hereditarily Indecomposable (H.I.) Banach spaces is provided. We apply this method to construct a nonseparable H.I. Banach space Y. This space is the dual, as well as the second dual, of a separable H.I. Banach space. Moreover the space of bounded linear operators $\mathcal{L}(Y)$ consists of elements of the form $\lambda I + W$ where W is a weakly compact operator and hence it has separable range. Another consequence of the exhibited method is the proof of the complete dichotomy for quotients of H.I. Banach spaces. Namely we show that every separable Banach space Z not containing an isomorphic copy of ℓ^1 is a quotient of a separable H.I. space X. Furthermore the isomorph of Z^* into X^*, defined by the conjugate operator of the quotient map, is a complemented subspace of X^*.

[1]Keywords: Quotient maps, Hereditarily Indecomposable Banach spaces, Nonseparable Banach spaces, operators on Banach spaces
[2]AMS subject classification[2000]: 46B20, 46B26, 46B03

Introduction

In this paper we present the construction of a nonseparable Hereditarily Indecomposable Banach space and we study its structure, as well as, the structure of the space of bounded linear operators acting on that space. Additionally we present a general method of producing Hereditarily Indecomposable Banach spaces. This method yields strong results concerning the quotients of H.I. Banach spaces. Among others, which will be explained later in the introduction, we obtain the complete dichotomy for quotients of H.I. spaces. Namely we prove that every separable Banach space Z, not containing isomorphically ℓ^1, is a quotient of a H.I. space X and moreover Z^* is isometric to a complemented subspace of X^*.

To explain our results in greater detail we begin by recalling the definition of Hereditarily Indecomposable (H.I.) Banach spaces. A Banach space X is H.I. provided that every infinite dimensional closed subspace admits only trivial projections. (A projection $P : X \longrightarrow X$ is called trivial if either $\operatorname{Im} P$ or $\operatorname{Ker} P$ is of finite dimension). The class of H.I. spaces introduced after the W.T. Gowers and B. Maurey example [**28**] and since then has been extensively studied by several researchers (see e.g. [**26**],[**25**],[**43**],[**20**],[**21**],[**34**]). As W.T. Gowers has proved, in his fundamental dichotomy [**27**], H.I. Banach spaces stand on the opposite side of Banach spaces with an unconditional basis. The precise statement of Gowers dichotomy is that every Banach space either contains a subspace with an unconditional basis or a H.I. subspace. Evidently, this result constitutes a major contribution towards the understanding of the fundamental prototypes of Banach spaces. Moreover H.I. Banach spaces derive a new class of spaces of the form $\mathcal{L}(X)$. Namely as it was shown in [**28**] every $T \in \mathcal{L}(X)$ (for a complex H.I. space X) has the form $T = \lambda I + S$ with S a strictly singular operator. The problem of the existence of Banach spaces X such that every $T \in \mathcal{L}(X)$ is of the form $T = \lambda I + K$ with K a compact operator remains open but if such a space exists it is natural to search for it in the class of H.I. spaces. Gowers and Maurey in [**28**] have presented general constructions of Banach spaces with small spaces of operators. Although these spaces are not always H.I. their methods are closely related to the original Gowers Maurey construction.

Recently it has been shown ([**5**], [**9**]) that the class of H.I. spaces is extensive. The main result contained in [**9**] is that every Banach space either contains isomorphically the space $\ell^1(\mathbb{N})$ or it has a subspace which is a quotient of a H.I. Banach space. This theorem yields that the dual space of a H.I. Banach space could contain isomorphically classical spaces like ℓ^p, $1 \leq p < \infty$. As we have mentioned above one of the main goals of the present paper is to prove the complete dichotomy concerning quotients of H.I. spaces.

---

[1]Research partially supported by Thales program of NTUA
[2]Received by the editor December 14, 2001

The first part of the paper mainly contains the construction of a nonseparable H.I. space. For those with experience in H.I. constructions it is clear that the methods used in the separable case are not extended to obtain nonseparable H.I. spaces. This happens for several reasons; one of them is the Gowers-Maurey coding function. Namely, in order to produce the special functionals which are responsible for the non existence of unconditional basic sequences in their space, Gowers and Maurey considered a certain injection of the set $\{(x_1, x_2, \ldots, x_n) : x_i \in c_{00}(\mathbb{N}),\ x_i(m) \in \mathbb{Q},\ x_i < x_{i+1}\}$ into the natural numbers. This is possible since the above set is countable and it is not easy to figure out what we should use in its place if we think of nonseparable constructions. Moreover there are classes of nonseparable Banach spaces which, as it is well known, do not contain any H.I. space. For example nonseparable weakly compactly generated Banach spaces (see [**3**]) or Banach spaces X with $\text{dens}(X) > \mathfrak{c}$ admit non trivial complemented subspaces. The first result contained in the present paper, and the proof of which follows from general properties of H.I. spaces, is that:

THEOREM 0.1. *Every Hereditarily Indecomposable Banach space X embeds into $\ell^\infty(\mathbb{N})$.*

This is a rather easy consequence of an alternative characterization of the H.I. property communicated to us by V.D. Milman.

The nonseparable H.I. space, presented here, is actually the dual as well as the second dual of a separable H.I. space. To explain how such a space is constructed we start with the next two observations. The first is the following:

If Y is a subspace of X such that the quotient map $Q : X \longrightarrow X/Y$ is strictly singular then B_Y is an "asymptotic" Kernel. That means that for every infinite dimensional subspace Z of X and every $\varepsilon > 0$ there exist $z \in S_Z$ and $y \in S_Y$ such that $\|z - y\| < \varepsilon$. If in addition Y is a H.I. space with a Schauder basis, the theory of bases (see e.g. [**36**]) easily yields that X is also H.I. The second is described by the next:

PROPOSITION 0.2. Let X be a Banach space with the properties:

(i) The space X does not contain ℓ^1.
(ii) The dual space X^* has a H.I. subspace Y such that the quotient space X^*/Y is isomorphic to $c_0(\Gamma)$ for some set Γ.

Then X^* is a H.I. space.

This result follows from the previous observation and the well known fact that assumption (i) and the fact $X^*/Y \cong c_0(\Gamma)$ imply that the quotient map $Q : X^* \longrightarrow X^*/Y$ is strictly singular [**16**]. Let's observe that if Γ is uncountable then X^* is nonseparable, hence we have shown the existence of nonseparable H.I. spaces. Moreover our H.I. constructions may be concerned with a separable subspace Y of X^*.

Let's state now the first of the main results of the present paper:

THEOREM 0.3. *There exists a Banach space X with a Schauder basis $(e_n)_{n\in\mathbb{N}}$ such that:*

(1) *The space X is an asymptotic ℓ^1 space, hence the basis $(e_n)_{n\in\mathbb{N}}$ is boundedly complete.*
(2) *The space X is Hereditarily Indecomposable and somewhat reflexive (i.e. every subspace of X contains a further reflexive subspace).*

(3) *The space* $X_* = \overline{\text{span}}\{e_n^* : n \in \mathbb{N}\}$ *is Hereditarily Indecomposable and somewhat reflexive.*
(4) *The quotient space* X^*/X_* *is isometric to* $c_0(\Gamma)$ *with* $|\Gamma| = \mathfrak{c}$, *hence* X^* *is nonseparable.*
(5) *The space* X^* *is Hereditarily Indecomposable and somewhat reflexive.*
(6) *The space* X^{**} *is isomorphic to* $X \oplus \ell^1(\Gamma)$.
(7) *The space* $\mathcal{L}(X)$ *of bounded linear operators on* X *has the form*

$$\mathcal{L}(X) = \{\lambda I + S : \ \lambda \in \mathbb{R}, \ S \text{ is strictly a singular and weakly compact operator }\}.$$

(8) *Every bounded linear operator* $T : X^* \longrightarrow X^*$ *is of the form* $T = S^* + K$ *where* $S \in \mathcal{L}(X)$ *and* K *is a compact operator.*
(9) *Every* $T \in \mathcal{L}(X^*)$ *is of the form* $T = \lambda I + R$ *for some operator* R *with separable range.*

Some of the properties listed above follow easily from others. For example property (5) is a consequence of the properties (3) and (4). It is also worth noticing that property (8) which is quite surprising appears as a consequence of the facts that X does not contain ℓ^1 and $X^{**} \cong X \oplus \ell^1(\Gamma)$. Hence this property remains valid in spaces not related to H.I. constructions like J. Hagler's space [**31**] and the predual of E. Odell's example of a nonseparable space not containing any subsymmetric basic sequence [**41**]. Examples of nonseparable Banach spaces X such that the space of operators $\mathcal{L}(X)$ are such as described in property (9) has been constructed by S. Shelah in [**48**] under the additional assumption $V = L$, and in ZFC by S. Shelah and J. Steprans [**49**]. A reflexive version of the previous examples has been also provided by H. Wark [**51**].

To explain briefly how such a space is obtained we first recall the main ingredients of the existing constructions of H.I. spaces like [**28**] and [**6**]. Any such construction is based on an unconditional frame which is either Schlumprecht's space [**47**] or the Argyros- Deliyanni asymptotic ℓ^1 mixed Tsirelson space [**6**]. Both are extensions of Tsirelson original construction of a Banach not containing c_0 or ℓ^p, $1 \le p < \infty$ ([**50**]). In the asymptotic ℓ^1 mixed Tsilerson case the space is defined as follows:

The space $T_{ad} = T[(\mathcal{S}_{n_j}, \frac{1}{m_j})_j]$ is the completion of $(c_{00}(\mathbb{N}), \|\ \|_*)$ where the norm $\|\ \|_*$ is defined by the following implicit formula:

$$\|x\|_* = \max\Big\{\|x\|_\infty, \ \sup_j \sup \frac{1}{m_j}\sum_{k=1}^{n}\|E_k x\|_*\Big\} \tag{1}$$

The inside supremum is taken over all families $(E_k)_{k=1}^n$ of finite subsets of the natural numbers such that $E_1 < E_2 < \cdots < E_n$ and $\{\min E_1, \min E_2, \dots, \min E_n\} \in \mathcal{S}_{n_j}$ ($\mathcal{S}_{n_j}$ denotes the n_j Schreier family).

Clearly this space has an unconditional basis. The next step in [**6**] is to use a Gowers Maurey coding function, mentioned above, in order to define a symmetric subset D of $B_{T_{ad}^*}$ which, as a norming set, yields a H.I. space. We denote such spaces $X = T[(\mathcal{S}_{n_j}, \frac{1}{m_j})_j, D]$.

One of the main ingredients defining the new spaces is described by the following:

DEFINITION 0.4. A subset G of c_{00} is said to be **ground**, provided that it contains the standard basis $(e_n)_{n\in\mathbb{N}}$ of c_{00}, it is symmetric and for $g \in G$ we have that $\|g\|_\infty \leq 1$, $g(n) \in \mathbb{Q}$ for all $n \in \mathbb{N}$, and $Eg \in G$ for every interval E of $\mathbb{N}$.

For a ground set G, the **ground norm** on c_{00} corresponding to G, is defined as $\|x\|_G = \sup\{< g, x >:\ g \in G\}$ and the **ground space** Y_G is the completion of the space $(c_{00}, \|\ \|_G)$.

Next we consider the space $T[G, (\mathcal{S}_{n_j}, \frac{1}{m_j})_j]$ with norm expressed by the implicit formula:

$$\|x\|_* = \max\Big\{\|x\|_G,\ \sup_j \sup \frac{1}{m_j} \sum_{k=1}^{n} \|E_k x\|_*\Big\} \tag{2}$$

where the conditions for the inside supremum are the same as in (1). Observe that for a general G the space $T[G, (\mathcal{S}_{n_j}, \frac{1}{m_j})_j]$ does not necessarily have an unconditional basis but it remains an asymptotic ℓ^1 space. The norm in [**6**] is defined with the ground set $G_0 = \{\pm e_n :\ n \in \mathbb{N}\}$. Next, similarly to [**6**] we choose a set D responsible for the H.I. structure. Its selection depends on the set G, uses a Gowers-Maurey coding function and D is a rationally convex and symmetric subset of the unit ball of $T^*[G, (\mathcal{S}_{n_j}, \frac{1}{m_j})_j]$. So we have arrived to a space with norming set D depended on two parameters G, $(\mathcal{S}_{n_j}, \frac{1}{m_j})_j$ and denoted by

$$X_G = T[G, (\mathcal{S}_{n_j}, \frac{1}{m_j})_j, D].$$

Is such a space H.I.? The answer in the general case is negative. For example setting $G = \{\pm\chi_F :\ F \text{ is a finite subset of } \mathbb{N}\}$ we easily observe that $\|\ \|_G$ is equivalent to the ℓ^1 norm hence the space X_G is isomorphic to $\ell^1(\mathbb{N})$. Therefore we need some more conditions on the set G. This is the property of G being $\mathcal{S}_2$ bounded, which is defined as follows:

DEFINITION 0.5. A ground set G is said to be $\mathcal{S}_2$ bounded, provided that the ground space Y_G satisfies the following two properties:

(i) The space $\ell^1(\mathbb{N})$ does not embed into Y_G.
(ii) For every weakly null block sequence $(y_n)_{n\in\mathbb{N}}$ in Y_G and $\varepsilon > 0$ there exists a subsequence $(y_n)_{n\in M}$ such that for every $g \in G$ the set

$$\{\min\operatorname{supp} y_n :\ n \in M \ \text{ and } |g(y_n)| > \varepsilon\}$$

belongs to $\mathcal{S}_2$, the second Schreier family.

For ground sets G which are $\mathcal{S}_2$ bounded we prove that:

THEOREM 0.6. *There exists a double sequence $(n_j, m_j)_{j\in\mathbb{N}}$ such that for every $\mathcal{S}_2$ bounded set G the space $X_G = T[G, (\mathcal{S}_{n_j}, \frac{1}{m_j})_j, D]$ has the following properties:*

(1) *The space X_G is asymptotic ℓ^1 and $(e_n)_{n\in\mathbb{N}}$ is a bimonotone boundedly complete Schauder basis.*
(2) *The space X_G is Hereditarily Indecomposable and somewhat reflexive.*
(3) *The space $(X_G)_* = \overline{\operatorname{span}}\{e_n^* :\ n \in \mathbb{N}\}$ is Hereditarily Indecomposable.*

(4) *The space* X_G^* *is the norm closed linear span of the closure of* G *in the topology of pointwise convergence i.e.* $X_G^* = \overline{\text{span}}(\overline{G}^p)$. *(*$\overline{G}^p$ *denotes the closure of* G *in the topology of pointwise convergence in* $\ell^\infty(\mathbb{N})$.*)*
(5) *The space of bounded linear operators of* X_G *has the form*

$$\mathcal{L}(X_G) = \{\lambda I + S : \ \lambda \in \mathbb{R}, \ \ S \textit{ is a strictly singular and weakly compact operator } \}.$$

Let's observe that, at least formally, this theorem yields a new variety of asymptotic ℓ^1 H.I. spaces. Examples of ground sets which are $\mathcal{S}_2$ bounded are those for which the ground space Y_G does not contain ℓ^1 and satisfies the weak Banach-Saks property. To explain briefly the role of $\mathcal{S}_2$ boundness of G we point out the following:

For a given normalized weakly null block sequence $(x_k)_{k\in\mathbb{N}}$ in the space X_G and given $\varepsilon > 0$ and $j > 1$, if $\dfrac{1}{m_j}$ is greater than ε then the "$\mathcal{S}_2$ bounded" property permits us to choose a certain convex combination $y = \sum a_k x_k$ ((ε, j) S.C.C. of a rapidly increasing sequence) such that the action of G on the vector y becomes negligible and hence we can use similar methods as in [**6**] to give an upper estimate of the norm of the vector y and to prove that the space X_G is H.I. The precise statement for these estimates is given in Chapter 4 where what we call the basic inequality is proved.

Let's pass now to see some concrete examples of $\mathcal{S}_2$ bounded sets G. All the examples we deal with are based on two tree structures on $\mathbb{N}$. Hence we denote by $\mathcal{D}$ the partial order of $\mathbb{N}$ as a binary tree. Furthermore we assume that the natural order of $\mathbb{N}$ coincides with the lexicographical order of $\mathcal{D}$. The second tree, denoted by $\mathcal{L}$, is defined as follows. Let $(\Lambda_i)_{i\in\mathbb{N}}$ be a partition of $\mathbb{N}$ into pairwise disjoint infinite sets. We define a partial order "$\preceq$" on $\mathbb{N}$ with the rule of $n \preceq m$ provided $n \leq m$ and $n, m \in \Lambda_i$ for some $i \in \mathbb{N}$. We also adopt the following notation. For s a finite segment of $\mathcal{D}$ (or $\mathcal{L}$) we denote by $s^* = \chi_s$ the characteristic function of s; s^* clearly belongs to $c_{00}(\mathbb{N})$. For s an infinite segment, s^* is similarly defined as an element of $\ell^\infty(\mathbb{N})$. Moreover by $\min s$ we denote the minimum of s in the natural order of $\mathbb{N}$. Consider the following set:

$$\begin{aligned} G_{NS} \;=\; & \Big\{ \sum_{i=1}^{d} \varepsilon_i s_i^* : \ (\varepsilon_i)_{i=1}^d \in \{-1,1\}^d \text{ and } (s_i)_{i=1}^d \text{ are pairwise disjoint} \\ & \text{finite segments of } \mathcal{D} \text{ and } \{\min s_1, \min s_2, \dots, \min s_d\} \in \mathcal{S}_1, \ \ d \in \mathbb{N} \Big\}. \end{aligned}$$

It is clear that G_{NS} is a ground set and it is shown that it is $\mathcal{S}_2$ bounded. The space $X_{G_{NS}} = T[G_{NS}, (\mathcal{S}_{n_j}, \frac{1}{m_j})_j, D]$ is the required space X of Theorem 0.3.

Consider next the following set:

$$\begin{aligned} G_0^{\mathcal{L}} \;=\; & \Big\{ \sum_{i=1}^{d} \varepsilon_i s_i^* : \ (s_i)_{i=1}^d \text{ are finite segments of } \mathcal{L}, \ s_i \subset \Lambda_i, \ \ (\varepsilon_i)_{i=1}^d \in \{-1,1\}^d \\ & \{\min s_1, \min s_2, \dots, \min s_d\} \in \mathcal{S}_1, \ \ d \in \mathbb{N} \Big\}. \end{aligned}$$

Then the space $X_{G_0^{\mathcal{L}}} = T[G_0^{\mathcal{L}}, (\mathcal{S}_{n_j}, \frac{1}{m_j})_j, D]$ satisfies all the properties of Theorem 0.3 with the only difference being that $X^*_{G_0^{\mathcal{L}}}/(X_{G_0^{\mathcal{L}}})_* = c_0(\mathbb{N})$ and $X^{**}_{G_0^{\mathcal{L}}} = X_{G_0^{\mathcal{L}}} \oplus \ell^1(\mathbb{N})$. Thus we obtain an example of a separable H.I. Banach space, that is $X^*_{G_0^{\mathcal{L}}}$, such that $\ell^1(\mathbb{N})$ is a complemented subspace of its dual.

The ℓ^2 variations of the above sets go as follows:

$$G_2^{\mathcal{D}} = \Big\{\sum_{i=1}^d a_i s_i^* : (s_i)_{i=1}^d \text{ are pairwise disjoint finite segments of } \mathcal{D}, \\ a_i \in \mathbb{Q}, \sum_{i=1}^d a_i^2 \leq 1, \ d \in \mathbb{N}\Big\}$$

The space $Y_{G_2^{\mathcal{D}}}$ is the well known James tree space ([**33**], [**35**]). For the space $X_{G_2^{\mathcal{D}}} = T[G_2^{\mathcal{D}}, (\mathcal{S}_{n_j}, \frac{1}{m_j})_j, D]$ the following holds:
For $(s_i)_{i=1}^d$, disjoint infinite segments of $\mathcal{D}$,

$$\|\sum_{i=1}^d a_i s_i^*\|_{X^*_{G_2^{\mathcal{D}}}} = (\sum_{i=1}^d a_i^2)^{\frac{1}{2}} \tag{3}$$

The reason for the above equality (3) is similar to a known property of T_{ad}. Namely the space c_0 is finitely representable in the subspaces of the space T_{ad} ([**7**],[**8**]). Equality (3) yields the following:

THEOREM 0.7. *There exists an asymptotic ℓ^1 H.I. space X with a boundedly complete Schauder basis $(e_n)_{n\in\mathbb{N}}$ such that*

(i) *X^*/X_* is isometric to $\ell^2(\mathfrak{c})$.*
(ii) *For every countable subset C of $\mathfrak{c}$ there exists a lifting of $\ell^2(C)$ into X^*. Hence X^* contains complemented subspaces isomorphic to ℓ^2.*

The set $G_2^{\mathcal{L}}$ is defined as:

$$G_2^{\mathcal{L}} = \Big\{\sum_{i=1}^d a_i s_i^* : (s_i)_{i=1}^d \text{ are finite segments of } \mathcal{L}, \ s_i \subset \Lambda_i, \\ a_i \in \mathbb{Q} \text{ for } i = 1, 2, \dots, d, \ \sum_{i=1}^d a_i^2 \leq 1, \ d \in \mathbb{N}\Big\}$$

The corresponding space $X_{G_2^{\mathcal{L}}} = T[G_2^{\mathcal{L}}, (\mathcal{S}_{n_j}, \frac{1}{m_j})_j, D]$ has the properties:

(a) $X_{G_2^{\mathcal{L}}}$ is asymptotic ℓ^1, H.I. space with a boundedly complete Schauder basis $(e_n)_{n\in\mathbb{N}}$.
(b) $(X_{G_2^{\mathcal{L}}})_*$ is H.I. and $X^*_{G_2^{\mathcal{L}}}$ is isomorphic to $(X_{G_2^{\mathcal{L}}})_* \oplus \ell^2(\mathbb{N})$.
(c) The correspondence $e_n \mapsto e_i$ iff $n \in \Lambda_i$ defines a quotient map $Q : X_{G_2^{\mathcal{L}}} \longrightarrow \ell^2$.

Similar results to the above are obtained for all $1 < p < \infty$. It is worthwhile to point out that beyond the complemented embedding of ℓ^2 we have a very specific quotient map described in (c). Much more general results hold; we will discuss them in the rest of the introduction.

All the above mentioned results are contained in chapters 1-11. We have tried to make the presentation self-contained and quite detailed. In chapters 12 we proceed to a detailed study of the structure of the Schreier families $(\mathcal{S}_\xi)_{\xi<\omega_1}$ and their modified versions. In chapter 13 we exhibit a general method of constructing H.I. spaces that extends the aforementioned results. Let's describe briefly the content of these methods.

When the construction of $T[(\mathcal{S}_{n_j}, \frac{1}{m_j}), D]$ appeared ([**6**]), it was known to the authors, but also clear to the experts, that for all $\xi < \omega_1$ similar results could be obtained for spaces of the form $T[(\mathcal{S}_{\xi_j}, \frac{1}{m_j}), D]$ with $(\xi_j)_{j\in\mathbb{N}}$ appropriate increasing sequence of ordinals and $\min_j \xi_j \geq \xi$. Therefore following exactly the same steps, as we have explained above, for every ground set G and increasing sequence $(\xi_j)_{j\in\mathbb{N}}$ we define the space $X_G = T[G, (\mathcal{S}_{\xi_j}, \frac{1}{m_j}), D]$. The question naturally raised, is under what conditions on G do we obtain a H.I. space? These conditions are described in the next.

Definition 0.8. A ground set G is said to be $\mathcal{S}_\xi$ bounded, for some $\xi < \omega_1$, provided the following two hold

(i) The ground space Y_G does not contain isomorphically ℓ^1.

(ii) For every weakly null block sequence $(y_n)_{n\in\mathbb{N}}$ in Y_G and every $\varepsilon > 0$ there exists an $L \in [\mathbb{N}]$ such that for every $g \in G$

$$\{\min \operatorname{supp} y_n : n \in L \text{ and } |g(y_n)| > \varepsilon\} \in \mathcal{S}_\xi.$$

The theorem for $\mathcal{S}_\xi$ bounded sets G is completely analogous to Theorem 0.6 stated before. Thus we have:

Theorem 0.9. *For every $\xi < \omega_1$ there exists a strictly increasing sequence $(\xi_j)_{j\in\mathbb{N}}$ of countable ordinals with $\xi_1 = \xi$, such that for every ground $\mathcal{S}_\xi$ bounded set G the space $X_G = T[G, (\mathcal{S}_{\xi_j}, \frac{1}{m_j})_j, D]$ has the following properties:*

(1) *The space X_G is asymptotic ℓ^1 and $(e_n)_{n\in\mathbb{N}}$ is a bimonotone boundedly complete Schauder basis.*

(2) *The space X_G is Hereditarily Indecomposable and somewhat reflexive.*

(3) *The space $(X_G)_* = \overline{\operatorname{span}}\{e_n^* : n \in \mathbb{N}\}$ is Hereditarily Indecomposable.*

(4) *The space X_G^* is the norm closed linear span of the closure of G in the topology of pointwise convergence i.e. $X_G^* = \overline{\operatorname{span}}(\overline{G}^p)$.*

(5) *The space of bounded linear operators of X_G has the form*

$\mathcal{L}(X_G) = \{\lambda I + S : \lambda \in \mathbb{R},\ S$ *is strictly a singular and weakly compact operator* $\}$.

The proof of this theorem follows exactly the steps of Theorem 0.6. Some small modifications are needed in some stages of the proof. The important advantage of the above theorem follows from the next result which is a consequence of the results of [**13**] (see also [**10**]).

Proposition 0.10. Let G be a ground set such that the ground space Y_G does not contain isomorphically ℓ^1. Then there exists $\xi < \omega_1$ such that G is $\mathcal{S}_\xi$ bounded.

A consequence is the following:

THEOREM 0.11. *Let Y be a Banach space not containing ℓ^1 and with a Schauder basis. Then for every normalized bimonotone Schauder basis $(y_n)_{n\in\mathbb{N}}$ of Y, there exists X an asymptotic ℓ^1 H.I. space, with a Schauder basis $(x_n)_{n\in\mathbb{N}}$ such that the correspondence $x_n \mapsto y_n$ extends to a bounded linear operator $T : X \longrightarrow Y$. Moreover if Y has separable dual or is reflexive, the corresponding property holds for the space X.*

The main application of Theorem 0.9 concerns quotients of H.I. spaces and is the following.

THEOREM 0.12. *Let Z be a separable Banach space not containing ℓ^1. Then there exists a H.I. Banach space X and a quotient map $Q : X \longrightarrow Z$ such that:*

(i) *X has a Schauder basis $(x_n)_{n\in\mathbb{N}}$ and it is an asymptotic ℓ^1 space.*
(ii) *$X^* = W \oplus Q^*Z^*$.*
(iii) *If Z has a Schauder basis $(z_k)_{k\in\mathbb{N}}$ then Q maps $(x_n)_{n\in\mathbb{N}}$ onto $(z_k)_{k\in\mathbb{N}}$ and also the subspace W of X^* is H.I.*
(iv) *If no basis assumption is made for Z then Q could chosen such that $\{Qx_n : n \in \mathbb{N}\}$ be a norm dense subset of the unit sphere of Z.*

Let's describe how this result is obtained. First we treat the case of Z with a Schauder basis. We choose a normalized bimonotone Schauder basis $(e_i)_{i\in\mathbb{N}}$ of Z and then, for the tree $\mathcal{L}$ mentioned before we consider the following set:

$$\begin{aligned} G_Z^{\mathcal{L}} = \Bigg\{ & E\sum_{i=1}^{d} a_i\Lambda_i^* : a_i \in \mathbb{Q} \text{ for } i = 1,2,\dots d, \\ & \sum_{i=1}^{d} a_i e_i^* \in B_{Z^*},\ d \in \mathbb{N},\ E \text{ is a finite interval of } \mathbb{N} \Bigg\}. \end{aligned}$$

where Eg denotes the restriction of the functional g on the interval E. Clearly $G_Z^{\mathcal{L}}$ is a ground set. It is shown that ℓ^1 does not embed into the ground space $Y_{G_Z^{\mathcal{L}}}$. Hence there exists $\xi < \omega_1$ such that $G_Z^{\mathcal{L}}$ is $\mathcal{S}_\xi$ bounded. Applying Theorem 0.9 we obtain the space $X_{G_Z^{\mathcal{L}}}$ which is H.I. and moreover Z^* is w^*−isometric to a complemented subspace of $X^*_{G_Z^{\mathcal{L}}}$. Furthermore $X^*_{G_Z^{\mathcal{L}}}$ is isomorphic to $(X_{G_Z^{\mathcal{L}}})_* \oplus Z^*$.

Now the case of a general space Z not containing ℓ^1 follows from the next:

THEOREM 0.13. *Let Z be a separable Banach space that contains no isomorphic copy of ℓ^1. Then there exists a space Y with a Schauder basis $(y_n)_{n\in\mathbb{N}}$ and a quotient map $Q : Y \longrightarrow Z$ such that ℓ^1 does not embed into Y and Q^*Z^* is a complemented subspace of Y^*.*

A corollary of Theorem 0.12 is the following, which is related to [**19**].

COROLLARY 0.14. Let $(X_i)_{i=1}^n$ be a finite family of reflexive H.I. Banach spaces. Assume that each X_i has a Schauder basis. Then there exists a H.I. Banach space Y such that $Y^* = W \oplus (\oplus_{i=1}^n X_i)$ where W is also H.I.

Also a consequence of the proof of Theorem 0.12 is the next result which provides quasi reflexive H.I. spaces. The problem of the existence of such spaces was posed to us by Gilles Godefroy.

THEOREM 0.15. *Let Z be a finite dimensional Banach space with a bimonotone basis. Then there exists an H.I. Banach space Y with the quotient space Y^{**}/Y isometric to Z. In particular for each $k \in \mathbb{N}$ there exists a quasi reflexive and H.I. space Y with* $\dim Y^{**}/Y = k$.

We should point out that Theorem 0.12 contains the complete dichotomy related to quotients of H.I. spaces. However does not cover completely the results of [**9**], since in the case of reflexive Z, the space X obtained in the present paper which has Z as a quotient is not reflexive. Additionally behind the results of [**9**] there also exists a general method of obtaining H.I. Banach spaces. The precise statement for this method is given in [**5**] and is the following:

THEOREM 0.16. *Let Z be a separable Banach space and W a bounded convex symmetric closed* **a**-*thin subset of Z. Then there exists a H.I. space X and a bounded injection $j : X \longrightarrow Z$ such that W is a subset of $j(B_X)$.*

This result is proved with interpolation methods; for a proof see also [**37**]. Let's point out the following difference between Theorems 0.12 and 0.16. Theorem 0.12 permits us to have a linear embedding of a ground set G such that the ground space Y_G does not contain ℓ^1 into the unit ball of the dual X^* for an appropriate H.I. space X, while Theorem 0.16 concerns the linear lifting of an **a**-thin set W in the ball of the H.I. space itself.

ACKNOWLEDGEMENT. We extend our warm thanks to E. Odell for his valuable help and his comments during the preparation of the present paper.

CHAPTER 1

General results about H.I. spaces

This first chapter contains some results obtained from the definition of H.I. Banach spaces and equivalent reformulations on it. Corollary 1.5 describes the frame on which we shall build the nonseparable H.I. space which will be presented in the next chapters. Also, Theorem 1.3 shows that every H.I. space is a subspace of $\ell^\infty(\mathbb{N})$.

The next Proposition summarizes the equivalent reformulations of the definition of H.I. Banach spaces.

PROPOSITION 1.1. Let X be a Banach space. The following assertions are equivalent:

(1) The space X is H.I.
(2) For every pair of infinite dimensional closed subspaces Y,Z of X, $\operatorname{dist}(S_Y, S_Z) = 0$.
(3) For every pair of infinite dimensional closed subspaces Y,Z of X and $\delta > 0$ there exist $y \in Y$ and $z \in Z$ such that $\|y - z\| \leq \delta \|y + z\|$.
(4) [V.D. Milman] For every infinite dimensional closed subspace Y of X, $\varepsilon > 0$ and $W \subset \mathrm{B}_{X^*}$ such that W ε-norms Y, the space $W_\perp = \{x \in X : f(x) = 0 \text{ for every } f \in W\}$ is a finite dimensional subspace of X.

Proof. The assertions (1),(2) and (3) are trivially equivalent by the open mapping theorem and the triangle inequality. We first show that (1) implies to (4). If Y is an infinite dimensional closed subspace of X, $\varepsilon > 0$ and $W \subset \mathrm{B}_{X^*}$ are such that W ε-norms Y then $Y \cap W_\perp = \{0\}$ and $Y \oplus W_\perp$ is a closed subspace of X. The H.I. property of X yields that $W_\perp$ is necessarily finite dimensional.

It remains to show that (4) implies (1). Suppose that the space X is not H.I. Then X has two infinite dimensional closed subspaces Y, Z such that $Y \cap Z = \{0\}$ and $Y + Z$ is a closed subspace of X. Then the projection $P : Y + Z \longrightarrow Y$ is continuous; set $\varepsilon = \frac{1}{\|P\|}$. For every $y^* \in \varepsilon \mathrm{B}_{Y^*}$ we may select, by Hahn-Banach theorem, a functional $\widetilde{y}^* \in \mathrm{B}_{X^*}$ such that $\widetilde{y}^*$ extends $y^* \circ P$ and $\|\widetilde{y}^*\|_{X^*} = \|y^* \circ P\|_{(Y+Z)^*}$. We set $W = \{\widetilde{y}^* : y^* \in \varepsilon \mathrm{B}_{Y^*}\}$. Then W ε-norms Y thus by (4) the space $W_\perp$ should be finite dimensional. This leads to a contradiction since clearly $W_\perp$ contains the infinite dimensional space Z. Therefore X is H.I. □

PROPOSITION 1.2. Let X be a H.I. Banach space, and $T : X \longrightarrow Y$ a bounded linear operator where Y is any Banach space. Then exactly one of the following two holds:

(i) The operator T is strictly singular.
(ii) $\operatorname{Ker} T$ is a finite dimensional subspace of X and X can be written as the direct sum of $\operatorname{Ker} T$ and a subspace Z of X such that $T_{|Z}$ is an isomorphism.

If $Y = X$ and T is one to one then either T is strictly singular or T is an onto isomorphism.

Proof. Assume that $\|T\| = 1$. Suppose that (i) does not hold. Then there exists an infinite dimensional closed subspace W of X and an $\varepsilon > 0$ such that $\|Tw\| \geq \varepsilon$ for every $w \in S_W$.

Suppose also that (ii) does not hold. Then the restriction of T to any subspace of X of finite codimension is not an isomorphism, thus, by Proposition 2.c.4 of [**36**] there exists an infinite dimensional subspace Z of X such that $\|T_{|Z}\| \leq \frac{\varepsilon}{2}$.

Then for every $z \in S_Z$ and $w \in S_W$ we have that

$$\|z - w\| \geq \|Tz - Tw\| \geq \|Tw\| - \|Tz\| \geq \varepsilon - \frac{\varepsilon}{2} = \frac{\varepsilon}{2}$$

so we get that $\text{dist}(S_Z, S_W) \geq \frac{\varepsilon}{2}$ which contradicts to the H.I. property of X. Hence $\operatorname{Ker} T$ is finite dimensional and has a complement Z such that $T_{|Z}$ is an isomorphism.

If $Y = X$ and T is one to one and not strictly singular then T is an isomorphism. By Corollary 19 and Theorem 21 of [**28**] a H.I. space is not isomorphic to any proper subspace, therefore T is onto. □

THEOREM 1.3. *Every H.I. Banach space X embeds into ℓ^∞.*

Proof. Let Y be any separable infinite dimensional closed subspace of X and select a countable subset D of the unit ball of X^* such that D $\frac{1}{2}$ norms Y. Since X is H.I. Proposition 1.1 yields that $D_\perp$ is finite dimensional. We enlarge D by a finite set F such that $(D \cup F)_\perp = \{0\}$. Let $D \cup F = \{x_n^* : n \in \mathbb{N}\}$. We define the operator $T : X \longrightarrow \ell^\infty$ by the rule $T(x) = (x_n^*(x))_{n\in\mathbb{N}}$. Observe that T is one to one and T restricted to Y is an isomorphism. Proposition 1.2 yields that T is an isomorphism. □

THEOREM 1.4. *Let X be a Banach space and let Z be an infinite dimensional closed subspace of X such that Z is H.I. and the quotient map $Q : X \longrightarrow X/Z$ is strictly singular. Then the space X is H.I.*

Proof. We begin the proof with the next two claims.

Claim 1. If Z_0 is a finite dimensional subspace of Z, Y is an infinite dimensional closed subspace of X and $\delta > 0$ then there exist $y \in S_Y$ and $z \in S_Z$ with $\|y - z\| < \delta$ such that $\|z_0\| < (1 + \delta)\|z_0 + \lambda z\|$ for every $z_0 \in Z_0$ and every scalar λ.

Proof of Claim 1. We may assume that $\delta < 1$. Let $\{x_1, x_2, \ldots, x_k\}$ be a $\frac{\delta}{4}$ net in S_{Z_0} and pick for each $i = 1, 2, \ldots, k$ an $f_i \in S_{X^*}$ with $f_i(x_i) = 1$. Then $\bigcap_{i=1}^k \operatorname{Ker} f_i$ is finite codimensional in X thus the subspace $Y \cap (\bigcap_{i=1}^k \operatorname{Ker} f_i)$ is infinite dimensional. From the fact that the operator $Q : X \longrightarrow X/Z$ is strictly singular, we may choose a $y \in Y \cap (\bigcap_{i=1}^k \operatorname{Ker} f_i)$ with $\|y\| = 1$ and $\|Qy\| < \frac{\delta}{16}$. If $z' \in Z$ with $\|y - z'\| < \frac{\delta}{16}$ then setting $z = \frac{z'}{\|z'\|}$ we get that $z \in S_Z$ and $\|y - z\| < \frac{\delta}{8}$.

To finish the proof of the claim it is enough to show that $\|z_0 + \lambda z\| > \frac{1}{1+\delta}$ for every $z_0 \in S_{Z_0}$ and every scalar λ. It is also enough to consider only λ with $|\lambda| < 2$.

Let $z_0 \in S_{Z_0}$ and $|\lambda| < 2$. We choose $i \in \{1, 2, \ldots, k\}$ such that $\|z_0 - x_i\| < \frac{\delta}{4}$. Then

$$\begin{aligned}
\|z_0 + \lambda z\| &\geq \|x_i + \lambda z\| - \|z_0 - x_i\| > |f_i(x_i + \lambda z)| - \frac{\delta}{4} \\
&\geq 1 - |\lambda| \cdot |f_i(z)| - \frac{\delta}{4} > 1 - |\lambda| \cdot (|f_i(y)| + \|y - z\|) - \frac{\delta}{4} \\
&> 1 - 2(0 + \frac{\delta}{8}) - \frac{\delta}{4} = 1 - \frac{\delta}{2} > \frac{1}{1+\delta}.
\end{aligned}$$

□

Claim 2. For every infinite dimensional closed subspace Y of X and every $\varepsilon > 0$ there exists an infinite dimensional closed subspace W of Z such that $\mathrm{dist}(w, S_Y) < \varepsilon$ for every $w \in S_W$.

Proof of Claim 2. Let $(\varepsilon_n)_{n \in \mathbb{N}}$ be a sequence of positive reals with $\prod_{n=1}^{\infty} (1 + \varepsilon_n) \leq 2$ and $\sum_{n=1}^{\infty} \varepsilon_n < \frac{\varepsilon}{8}$. From Claim 1 we may inductively select a sequence $(y_n)_{n \in \mathbb{N}}$ in S_Y and a sequence $(z_n)_{n \in \mathbb{N}}$ in S_Z such that $\|y_n - z_n\| < \varepsilon_n$ and $\|\sum_{i=1}^{n} a_i z_i\| < (1 + \varepsilon_{n+1}) \|\sum_{i=1}^{n+1} a_i z_i\|$ for every choice of scalars $(a_i)_{i \in \mathbb{N}}$ and all n. Then $(z_n)_{n \in \mathbb{N}}$ is a Schauder basic sequence with basis constant less or equal to 2. We set $W = \overline{\mathrm{span}}\{z_n : n \in \mathbb{N}\}$.

The space W satisfies the conclusion of the claim. Indeed, let $w \in S_W$, $w = \sum_{n=1}^{\infty} a_n z_n$. Then $|a_n| \leq 4$ for all n. The series $\sum_{n=1}^{\infty} a_n y_n$ converges to some $y' \in Y$ and $\|w - y'\| \leq \sum_{n=1}^{\infty} |a_n| \|y_n - z_n\| \leq 4 \sum_{n=1}^{\infty} \varepsilon_n < \frac{\varepsilon}{2}$. Setting $y = \frac{y'}{\|y'\|}$ we obtain that

$$\mathrm{dist}(w, S_Y) \leq \|w - y\| \leq \|w - y'\| + \|y' - y\| \leq \frac{\varepsilon}{2} + |\, \|y'\| - 1| < \varepsilon.$$

□

We pass now to the proof that the space X is H.I. Let Y_1, Y_2 be a pair of infinite dimensional closed subspaces of X^* and we will show that $\mathrm{dist}(S_{Y_1}, S_{Y_2}) = 0$.

Let $\varepsilon > 0$. From Claim 2, there exist two infinite dimensional closed subspaces W_1, W_2 of Z such that for every $w_1 \in S_{W_1}$ we have that $\mathrm{dist}(w_1, S_{Y_1}) < \frac{\varepsilon}{3}$ and for every $w_2 \in S_{W_2}$ we have that $\mathrm{dist}(w_2, S_{Y_2}) < \frac{\varepsilon}{3}$. Since the space Z is H.I. we may choose $w_1 \in S_{W_1}$ and $w_2 \in S_{W_2}$ with $\|w_1 - w_2\| < \frac{\varepsilon}{3}$. Let $y_1 \in S_{Y_1}$ such that $\|w_1 - y_1\| < \frac{\varepsilon}{3}$ and $w_2 \in S_{W_1}$ such that $\|w_2 - y_2\| < \frac{\varepsilon}{3}$. We deduce that

$$\begin{aligned}
\mathrm{dist}(S_{Y_1}, S_{Y_2}) &\leq \|y_1 - y_2\| \\
&\leq \|y_1 - z_1\| + \|z_1 - z_2\| + \|z_2 - y_2\| \\
&< \frac{\varepsilon}{3} + \frac{\varepsilon}{3} + \frac{\varepsilon}{3} = \varepsilon.
\end{aligned}$$

Thus $\mathrm{dist}(S_{Y_1}, S_{Y_2}) = 0$. Therefore the space X is H.I. □

With the next result we provide some sufficient conditions yielding the H.I. property for dual Banach spaces.

COROLLARY 1.5. Let X be a Banach space and let Z a subspace of X^* such that the following conditions are fulfilled.

(i) The space X contains no isomorphic copy of ℓ^1.
(ii) The space Z is H.I.
(iii) The space X^*/Z is isomorphic to $c_0(\Gamma)$ for some set Γ.

Then the space X^* is H.I.

Proof. From Theorem 1.4 it is enough to observe that the quotient map $Q : X^* \longrightarrow X^*/Z$ is strictly singular.

If Γ is finite there is nothing to be proved. Let Γ be infinite and suppose that Q is not strictly singular. Let Y be an infinite dimensional subspace of X such that Q restricted to Y is an isomorphism. Since X^*/Z is isomorphic to $c_0(\Gamma)$ for an infinite set Γ we may assume, passing to a subspace, that $Q(Y)$ is isomorphic to c_0. It follows that X^* contains isomorphically c_0 thus, by Bessaga Pelczynski's Theorem [**16**], the space X contains isomorphically ℓ^1, a contradiction. □

Similar arguments also yield the next.

COROLLARY 1.6. Let X be a Banach space and Y a closed subspace of X. If the spaces Y and X/Y are H.I. saturated then the same holds for the space X.

A similar result also holds for somewhat reflexive (or reflexive saturated) Banach spaces. We close this chapter with the following result.

PROPOSITION 1.7. Let X be a Banach space such that

(i) X does not contain isomorphically ℓ^1.
(ii) X^{**} is isomorphic to $X \oplus \ell^1(\Gamma)$ for some infinite set Γ.

Then every bounded linear operator $T : X^* \longrightarrow X^*$ is of the form $T = Q^* + K$ where Q is an operator on X and K is a compact operator on X^*.

Proof. Let $T : X^* \longrightarrow X^*$ be a bounded linear operator. Consider the conjugate operator $T^* : X^{**} \longrightarrow X^{**}$ and the projections

$$P_1 : X \oplus \ell^1(\Gamma) \longrightarrow X \text{ and } P_2 : X \oplus \ell^1(\Gamma) \longrightarrow \ell^1(\Gamma).$$

Let $Q : X \longrightarrow X$ be the operator defined as $Q = P_1 \circ (T^*|X)$ and $S : X \longrightarrow X^{**}$ defined by $S = P_2 \circ (T^*|X)$. Observe that $S(X) \subset \ell^1(\Gamma)$.

The operator S is compact. Indeed, if S was not compact we could construct an operator from a subspace of X onto ℓ^1, thus by the lifting property of ℓ^1, X contains isomorphically ℓ^1, a contradiction. It follows that the operator $S^* : X^{***} \longrightarrow X^*$ is compact so the same holds for the operator $K = S^*|X^*$. An easy computation yields that $T = Q^* + K$. □

REMARK 1.8. Hagler's space [**31**] and the predual of Odell's space [**41**], satisfy conditions (i) and (ii), and thus the conclusion of Proposition 1.7.

CHAPTER 2

Schreier families and repeated averages

The H.I. constructions, contained in this paper, require the use of Schreier families and the repeated averages as they developed in [**1**], [**6**],[**13**],[**10**]. This chapter is mainly devoted to recalling the basic definitions and to prove some estimates necessary in the sequel. We concentrate on the sequence $(\mathcal{S}_n)_{n\in\mathbb{N}}$ and the corresponding repeated averages $\{a(n,L) : n \in \mathbb{N}, L \in [\mathbb{N}]\}$ which we call $n-$averages. This material is sufficient to present the H.I. constructions for ground sets G which are $\mathcal{S}_2$ bounded. In chapter 12 we discuss corresponding definitions and results in the general setting.

For A, B finite subsets of the natural numbers we write $A < B$ if either at least one of the sets A, B is empty or A, B are nonempty and $\max A < \min B$. In this case A, B are called *successive*. For $n \in \mathbb{N}$ we write $n < A$ instead of $\{n\} < A$ and $A < n$ instead of $A < \{n\}$. We also write $n \leq A$ if $n \leq \min A$ or A is empty. We denote by $\#A$ the cardinality of a finite set A. A finite interval of the natural numbers is any set of the form $\{i \in \mathbb{N} : k \leq i \leq l\}$ for $k, l \in \mathbb{N}$ with $k \leq l$ while an infinite interval is any set of the form $\{i \in \mathbb{N} : k \leq i\}$ for $k \in \mathbb{N}$. c_{00} denotes the vector space of all eventually zero sequences of reals and $e_1, e_2, e_3, \ldots$ denotes the standard Hamel basis of c_{00}. For $x = \sum\limits_{i=1}^{\infty} a_i e_i \in c_{00}$ the support of x, denoted by $\mathrm{supp}(x)$, is the finite set $\{i \in \mathbb{N} : a_i \neq 0\}$ while $\mathrm{ran}(x)$ is the smallest interval of the natural numbers that contains $\mathrm{supp}(x)$. For a set E and $x = \sum\limits_{i=1}^{\infty} a_i e_i \in c_{00}$ we denote by Ex the vector $Ex = \sum\limits_{i\in E} a_i e_i$. For $x, y \in c_{00}$, $x = \sum\limits_{i=1}^{\infty} a_i e_i$, $y = \sum\limits_{i=1}^{\infty} b_i e_i$ we denote by $y(x)$ (or by $\langle y, x\rangle$) the inner product $y(x) = \langle y, x\rangle = \sum\limits_{i=1}^{\infty} a_i b_i$. The same notation remains valid if one of the vectors x, y belongs to c_{00} and the other to l^∞.

A family $\mathcal{M}$ of finite subsets of $\mathbb{N}$ is called

(1) *Compact* if the set $\{\chi_A : A \in \mathcal{M}\}$ is a compact subset of $\{0,1\}^{\mathbb{N}}$.
(2) *Hereditary* if for every $A \in \mathcal{M}$ and $B \subset A$ we have $B \in \mathcal{M}$.
(3) *Spreading* if for every $A = \{t_1 < t_2 < \cdots < t_r\} \in \mathcal{M}$ and $B = \{t'_1 < t'_2 < \cdots < t'_r\}$ with $t_i \leq t'_i$ for $i = 1, 2, \ldots, r$ we have $B \in \mathcal{M}$.

Let $\mathcal{M}$ be a family of finite subsets of $\mathbb{N}$. A finite sequence $(E_1, E_2, \ldots, E_n)$ of successive subsets of $\mathbb{N}$ is said to be $\mathcal{M}$-*admissible* if there exists a set $\{m_1 < m_2 < \cdots < m_n\} \in \mathcal{M}$ such that $m_1 \leq E_1 < m_2 \leq E_2 < \cdots < m_n \leq E_n$. A finite sequence $(f_1, f_2, \ldots, f_n)$ of vectors in c_{00} is said to be $\mathcal{M}$-*admissible* if $(\mathrm{supp}\, f_1, \mathrm{supp}\, f_2, \ldots, \mathrm{supp}\, f_n)$ is $\mathcal{M}$-admissible.

For $\mathcal{M}, \mathcal{N}$ families of finite subsets of $\mathbb{N}$ the convolution of $\mathcal{M}$ and $\mathcal{N}$, introduced in [**6**], is defined as follows

$$\begin{aligned}\mathcal{M} * \mathcal{N} \;=\; & \Big\{F \subset \mathbb{N} : \text{ there exists } s \in \mathbb{N},\ F_1, F_2, \ldots, F_s \in \mathcal{N} \text{ and} \\ & \{m_1 < m_2 < \cdots < m_s\} \in \mathcal{M} \text{ such that } m_1 \leq F_1 < m_2 \leq F_2 \\ & < \cdots < m_s \leq F_s \text{ and } \quad F = \bigcup_{i=1}^{s} F_i\Big\} \cup \Big\{\emptyset\Big\}.\end{aligned}$$

It can be easily seen that $*$ is associative in the set of all families of finite sets of $\mathbb{N}$ and that if $\mathcal{M}, \mathcal{N}$ are compact, hereditary and spreading the same holds for $\mathcal{M} * \mathcal{N}$.

The first Schreier family is the following:

$$\mathcal{S} = \{F \subset \mathbb{N} : \#F \leq \min F\} \cup \{\emptyset\}$$

and it is easy to see that is a compact, hereditary and spreading family.

The Schreier families $(\mathcal{S}_\xi)_{\xi<\omega_1}$ are defined as follows:

$$\mathcal{S}_0 = \Big\{\{t\} : \quad t \in \mathbb{N}\Big\} \cup \{\emptyset\}.$$

Let $\xi < \omega_1$ and suppose that $\mathcal{S}_\zeta$ has been defined for all $\zeta < \xi$. If $\xi = \zeta + 1$ we set

$$\mathcal{S}_\xi = \mathcal{S} * \mathcal{S}_\zeta.$$

If ξ is a limit ordinal fix a strictly increasing sequence $(\xi_n)_{n\in\mathbb{N}}$ of non limit ordinals with $\sup\limits_n \xi_n = \xi$ and set

$$\mathcal{S}_\xi = \bigcup_{n=1}^{\infty} \{F \in \mathcal{S}_{\xi_n} : n \leq F\}.$$

Each $\mathcal{S}_\xi$ is, as can be easily verified by transfinite induction, compact, hereditary and spreading. Another property of the Schreier families is that if $\zeta < \xi < \omega_1$ then there exists a $d = d(\zeta, \xi) \in \mathbb{N}$ such that every member F of $\mathcal{S}_\zeta$ with $F \geq d$ belongs also to $\mathcal{S}_\xi$.

The modified Schreier families are defined recursively as follows:

$$\begin{aligned}\mathcal{S}_0^M \;=\; & \Big\{\{t\} : \quad t \in \mathbb{N}\Big\} \cup \{\emptyset\} \\ \mathcal{S}_{\xi+1}^M \;=\; & \Big\{F \subset \mathbb{N} : \text{ there exists } k \in \mathbb{N} \text{ and } F_1, F_2, \ldots, F_k \in \mathcal{S}_\xi^M \\ & \quad \text{such that } k \leq F_i \text{ for } i = 1, 2, \ldots, k \text{ and } F = \bigcup_{i=1}^{k} F_i\Big\}\end{aligned}$$

and if $\xi < \omega_1$ is a limit ordinal and $(\xi_n)_{n\in\mathbb{N}}$ is the strictly increasing sequence of ordinals that defines $\mathcal{S}_\xi$ we set

$$\mathcal{S}_\xi^M = \bigcup_{n=1}^{\infty} \{F \in \mathcal{S}_{\xi_n}^M : n \leq F\}$$

It is easy to see by transfinite induction that each $\mathcal{S}_\xi^M$ is compact hereditary and spreading and that $\mathcal{S}_\xi \subset \mathcal{S}_\xi^M$. As it is proved in [**7**]:

PROPOSITION 2.1. *For every* $n < \omega$, $\mathcal{S}_n = \mathcal{S}_n^M$.

It is not clear if the above equality holds for all $\xi < \omega_1$.

Definition of $\{a(n,L) : L \in [\mathbb{N}], \ n \in \mathbb{N}\}$.

The $n-$averages $a(n,L)$ are defined as elements of c_{00} in the following manner:
For $n = 0$ we set $a(0,L) = e_{l_1}$ where $l_1 = \min L$.
Suppose that $a(n,L)$ has been defined for some $n \in \mathbb{N}$ and every $L \in [\mathbb{N}]$. Let $L \in [\mathbb{N}]$. We define $a(n+1,L)$ in the following way:
We set $l_1 = \min L$, $L_1 = L$ and $L_k = L_{k-1} \setminus \operatorname{supp} a(n, L_{k-1})$ for $k = 2, \dots, l_1$. Then we define

$$a(n+1,L) = \frac{1}{l_1}(a(n,L_1) + a(n,L_2) + \cdots + a(n,L_{l_1}))$$

Our notation is someway different from that used in [**13**] where the family $\{\xi_n^L : L \in [\mathbb{N}], \ n \in \mathbb{N}\}$ of repeated averages introduced. In our notation $a(n,L)$ stands for $(n)_1^L$ in the terminology of [**13**].

REMARK 2.2. The following properties of the $n-$averages $\{a(n,L) : L \in [\mathbb{N}], \ n \in \mathbb{N}\}$ are easily established by induction:

(1) $a(n,L)$ is a convex combination of the standard basis $(e_i)_{i \in \mathbb{N}}$ of c_{00}.
(2) $a(n,L) \in S_{\ell^1}^+$ i.e. $\|a(n,L)\|_{\ell^1} = 1$ and $a(n,L)(k) \geq 0$ for all $k \in \mathbb{N}$.
(3) $\operatorname{supp} a(n,L)$ is the maximal initial segment of L contained in $\mathcal{S}_n$.
(4) $\|a(n,L)\|_\infty = \dfrac{1}{l_1^n}$ where $l_1 = \min L$.
(5) If $\operatorname{supp} a(n,L) = \{t_1 < t_2 < \cdots < t_d\}$ and $a(n,L) = \sum_{i=1}^{d} a_{t_i} e_{t_i}$ then $a_{t_1} \geq a_{t_2} \geq \cdots \geq a_{t_d}$.

PROPOSITION 2.3. Let $n \in \mathbb{N}$, $n \neq 0$ and $L \in [\mathbb{N}]$. Then for every $G \in \mathcal{S}_{n-1}$ we have that

$$\sum_{k \in G} a(n,L)(k) < \frac{3}{l_1}$$

where $l_1 = \min L$.

Proof. For $f \in c_{00}$ and $G \subset \mathbb{N}$ we use the notation $\prec f, G \succ = \sum_{k \in G} f(k)$. We may assume, without loss of generality, that $G \subset L$, G is a maximal $\mathcal{S}_{n-1}$ set and also that $l_1 = \min L \geq 3$.
The proof is inductive (on n). For technical reasons we initially make the first two inductive steps, namely for $n = 1$ and $n = 2$.
For $n = 1$. Let $L \in [\mathbb{N}]$. Then G, as an $\mathcal{S}_0$ maximal set is a singleton. Thus

$$\sum_{k \in G} a(1,L)(k) \leq \|a(1,L)\|_\infty = \frac{1}{l_1} \quad (< \frac{3}{l_1}).$$

For $n = 2$. Let $L \in [\mathbb{N}]$. We consider the sets

$$L_1 = L, \quad L_{j+1} = L_j \setminus \operatorname{supp} a(1, L_j) \text{ for } j = 1, 2, \dots, l_1 - 1$$

where $l_1 = \min L$. Setting $l_j = \min L_j = \min \operatorname{supp} a(1, L_j)$ for $j = 1, 2, \dots, l_1$ we observe that $l_j \geq 2 l_{j-1}$ for all $j = 2, \dots, l_1$ since $\operatorname{supp} a(1, L_j)$ are $\mathcal{S}_1$ maximal sets. Let j_0 be the unique j such that $\min G \in \operatorname{supp} a(1, L_j)$. Since G is $\mathcal{S}_1$ maximal, setting $d = \min G$ we have that $\#G = d$. We also observe that $l_{j_0+2} \geq 2 l_{j_0+1} > 2d$.

These yield the following:

$$\sum_{k\in G} a(2,L)(k) = \prec a(2,L), G \succ = \frac{1}{l_1}\sum_{j=j_0}^{l_1} \prec a(1,L_j), G \succ$$

$$= \frac{1}{l_1}\left(\prec a(1,L_{j_0}), G \succ + \prec a(1,L_{j_0+1}), G \succ + \sum_{j=j_0+2}^{l_1} \prec a(1,L_j), G \succ \right)$$

$$\leq \frac{1}{l_1}\left(1+1+\sum_{j=j_0+2}^{l_1}\frac{d}{l_j}\right) < \frac{1}{l_1}\left(1+1+d\frac{2}{l_{j_0+2}}\right)$$

$$< \frac{1}{l_1}(1+1+1) = \frac{3}{l_1}.$$

For the general inductive step suppose that the conclusion holds for $n(\geq 2)$ and we show it for $n+1$. Let $L \in [\mathbb{N}]$. We set $l_1 = \min L$ and consider the sets $L_1, \ldots, L_{l_1}$ as in the definition. Setting $l_j = \min L_j = \min \operatorname{supp} a(n, L_j)$ for $j = 1, 2, \ldots, l_1$ we observe that for each $j = 2, \ldots, l_1$ we have that:

$$\frac{l_{j+1}}{l_j} = \frac{\min\operatorname{supp} a(n,L_{j+1})}{l_j} \geq \frac{\max\operatorname{supp} a(n,L_j)\ +1}{l_j} \geq \frac{\max\operatorname{supp} a(2,L_j)\ +1}{l_j}$$

$$\geq \frac{2^{l_j} l_j}{l_j} = 2^{l_j} \geq 2^3 = 8. \tag{4}$$

(Inequality $\max\operatorname{supp} a(2,L_j)+1 \geq 2^{l_j} l_j$ holds since for each j the set $\operatorname{supp} a(2,L_j)$ is $\mathcal{S}_2$ maximal so it is the union of l_j successive maximal $\mathcal{S}_1$ sets $L_j^1 < L_j^2 < \cdots < L_j^{l_j}$ and for each $k = 2, \ldots, l_j$ we have that $\min L_j^k \ \geq\ 2\min L_j^{k-1}$).

Let j_0 be the unique j such that $\min G \in \operatorname{supp} a(n, L_j)$. Since G is $\mathcal{S}_n$ maximal, setting $d = \min G$ there exist $G_1, G_2, \ldots, G_d$ maximal $\mathcal{S}_{n-1}$ sets such that $G_1 < G_2 < \cdots < G_d$ and $G = \bigcup_{i=1}^{d} G_i$. From (4) we have that $l_{j_0+2} \geq 8 l_{j_0+1} > 8d$. Hence the last relation and the inductive hypothesis yield that:

$$\sum_{k\in G} a(n+1,L)(k) = \prec a(n+1,L), G \succ\ = \frac{1}{l_1}\sum_{j=j_0}^{l_1} \prec a(n,L_j), G \succ$$

$$= \frac{1}{l_1}\left(\prec a(n,L_{j_0}), G \succ + \prec a(n,L_{j_0+1}), G \succ + \sum_{j=j_0+2}^{l_1} \prec a(n,L_j), G \succ \right)$$

$$\leq \frac{1}{l_1}\left(1+1+\sum_{j=j_0+2}^{l_1}\sum_{i=1}^{d} \prec a(n,L_j), G_i \succ\right) < \frac{1}{l_1}\left(1+1+\sum_{j=j_0+2}^{l_1}\frac{3d}{l_j}\right)$$

$$< \frac{1}{l_1}\left(1+1+3d\frac{8}{7l_{j_0+2}}\right) < \frac{1}{l_1}\left(1+1+\frac{3}{7}\right) < \frac{3}{l_1}.$$

□

NOTATION 2.4. Let $\vec{x} = (x_k)_{k\in\mathbb{N}}$ be a block sequence in c_{00}. Setting $t_k = \min\operatorname{supp} x_k$ for $k = 1, 2, \ldots$ and $L = \{t_1, t_2, \ldots\}$ we use the notation:

$$a(n, \vec{x}) = a(n, (x_k)_{k\in\mathbb{N}}) = \sum_k a(n, L)(t_k) x_k.$$

CHAPTER 3

The space $X = T\left[G, \left(\mathcal{S}_{n_j}, \frac{1}{m_j}\right)_j, D\right]$ and the auxiliary space T_{ad}

The definition of the space $X = T\left[G, \left(\mathcal{S}_{n_j}, \frac{1}{m_j}\right)_j, D\right]$ requires certain successive steps which we present in this chapter. We should point out that, although we follow closely the definition of asymptotic ℓ^1 spaces presented in [**6**], our presentation here has some differences. First, of course, is the ground set G appearing from the very beginning. Also the set D, as it is defined here, does not contain yet, the special functionals. Moreover, the familiar reader will observe that we close in each step of the inductive definition the set W in rational convex combinations. This makes our computations a little more complicated, but it seems necessary if we want to study the H.I. property of the predual X_* of X, as well as, the structure of the space of operators $\mathcal{L}(X)$. We also present the auxiliary space T_{ad} and we estimate the action of functionals on special convex combinations of the basis T_{ad}. These, as in [**6**], will help us to compute the norm of certain special convex combinations in the space X.

DEFINITION 3.1. A set $G \subset c_{00}$ is said to be *ground* provided the following conditions are fulfilled:

(1) $e_n \in G$ for all $n = 1, 2, \ldots$
(2) $g(n) \in \mathbb{Q}$ for every $g \in G$ and $n \in \mathbb{N}$.
(3) G is symmetric.
(4) For every $g \in G$ $\|g\|_\infty \leq 1$.
(5) For every $g \in G$ and E interval of $\mathbb{N}$ we have $Eg \in G$.

The second assertion yields that G is a countable set.

DEFINITION 3.2. Let G be a ground set. The *ground norm* on c_{00} corresponding to G is defined on c_{00} as $\|x\|_G = \sup\{g(x) : g \in G\}$ and the *ground space* Y_G is defined as the completion of the space $(c_{00}, \| \, \|_G)$.

Observe that the definition of the ground set yields that the standard basis $(e_n)_{n\in\mathbb{N}}$ of c_{00} defines a normalized bimonotone Schauder basis of the space Y_G. Hence also $B^*_{Y_G} = \overline{\text{conv}(G)}^p$.

For a set $B \subset c_{00}$ we denote by $\text{conv}_{\mathbb{Q}}(B)$ the set of all convex combinations of B with rational coefficients. We are ready to define the set $W = W\left[G, (\mathcal{S}_{n_j}, \frac{1}{m_j})_j\right]$. The norming sets of the spaces we are interested for will be defined as subsets of the corresponding sets W.

DEFINITION 3.3. Let $(m_j)_{j\in\mathbb{N}}$,$(n_j)_{j\in\mathbb{N}}$ be two strictly increasing sequences of natural numbers with $m_1 = 2$. We recursively define two sequences $(A_n)_{n\in\mathbb{N}}$ and $(W_n)_{n\in\mathbb{N}}$ of subsets of c_{00} as follows:

$$A_0 = G, \qquad W_0 = \mathrm{conv}_{\mathbb{Q}}(G).$$

If A_n and W_n have been defined we set:

$$\begin{aligned} A_{n+1} &= \Big\{\frac{1}{m_j}\sum_{k=1}^{d} f_k : \ f_k \in W_n \text{ for } k=1,2,\dots,d \text{ and } (f_1,f_2,\dots,f_d) \\ &\quad \text{is an } \mathcal{S}_{n_j} \text{ admissible sequence}, j\in\mathbb{N}\Big\}\bigcup A_n \\ W_{n+1} &= \mathrm{conv}_{\mathbb{Q}}(A_{n+1}). \end{aligned}$$

We set

$$W = \bigcup_{n=0}^{\infty} W_n \text{ and } A = \bigcup_{n=0}^{\infty} A_n.$$

The following properties are easily established:

(a) $e_n \in W$ for each $n = 1,2,\dots$

(b) The set W is countable, symmetric and closed under rational convex combinations.

(c) For every $f \in W$ and E interval of $\mathbb{N}$ we have that $Ef \in W$.

(d) For every $f \in W$ we have $\|f\|_\infty \le 1$. Additionally if f is of the form

$$f = \frac{1}{m_j}(f_1 + f_2 + \cdots + f_d)$$

for an $\mathcal{S}_{n_j}$ admissible sequence $(f_1,f_2,\dots,f_d)$ in W then $\|f\|_\infty \le \dfrac{1}{m_j}$.

The following proposition describes the norm defined by the set W and is proved in a similar manner as the corresponding result in the classical Tsirelson space ([**22**]).

PROPOSITION 3.4. Let $\|\ \|_W$ be the norm defined on c_{00} as $\|x\|_W = \sup\{f(x) : f \in W\}$. Then the norm $\|\ \|_W$ satisfies the following implicit formula:

$$\|x\|_W = \max\Big\{\|x\|_G,\ \sup_j \sup \frac{1}{m_j}\sum_{k=1}^{n} \|E_k x\|_W\Big\}$$

where the inside supremum is taken over all families $(E_k)_{k=1}^n$ of finite successive subsets of the natural numbers such that the family $(E_1,E_2,\dots,E_n)$ is $\mathcal{S}_{n_j}$ admissible.

DEFINITION 3.5 (The tree $\boldsymbol{T_f}$ of a functional $\boldsymbol{f \in W}$). Let $f \in W$. By a *tree* of f (or tree corresponding to the analysis of f) we mean a finite family $T_f = (f_a)_{a\in\mathcal{A}}$ indexed by a finite tree $\mathcal{A}$ with a unique root $0 \in \mathcal{A}$ such that the following conditions are satisfied:

1. $f_0 = f$ and $f_a \in W$ for all $a \in \mathcal{A}$.
2. An $a \in \mathcal{A}$ is maximal if and only if $f_a \in G$.
3. For every $a \in \mathcal{A}$ which is not maximal, denoting by S_a the set of the immediate successors of a, exactly one of the following two holds:

(a) $S_a = \{\beta_1, \dots, \beta_d\}$ with $f_{\beta_1} < \cdots < f_{\beta_d}$ and there exists $j \in \mathbb{N}$ such that the family $(f_{\beta_1}, \dots, f_{\beta_d})$ is $\mathcal{S}_{n_j}$ admissible and $f_a = \frac{1}{m_j} \sum_{i=1}^{d} f_{\beta_i}$

(b) $S_a = \{\beta_1, \dots, \beta_d\}$ and there exists a family of positive rationals $\{r_{\beta_i} : i = 1, \dots, d\}$ with $\sum_{i=1}^{d} r_{\beta_i} = 1$ such that $f_a = \sum_{i=1}^{d} r_{\beta_i} f_{\beta_i}$. Moreover for all $i = 1, \dots, d$, $\operatorname{ran} f_{\beta_i} \subset \operatorname{ran} f_a$.

It is obvious that every $f \in W$ has a tree which is not necessarily unique.

DEFINITION 3.6 (Types of functionals). Let $f \in W$ and let $T_f = (f_a)_{a \in \mathcal{A}}$ be a tree of f. For $a \in \mathcal{A}$ the functional f_a is said to be of type I if the set S_a satisfies condition (3a) in the above definition while f_a is said to be of type II if S_a satisfies the corresponding condition (3b).

DEFINITION 3.7 (The weight $\boldsymbol{w(f)}$ of an $\boldsymbol{f \in A}$). For an $f \in A$ we denote by $w(f)$ the weight of f which is 1 if $f \in G$ and m_j if f is of the form $f = \frac{1}{m_j}(f_1 + f_2 + \cdots + f_d)$ where $(f_1, f_2, \dots, f_d)$ is an $\mathcal{S}_{n_j}$ admissible sequence in W.

The norming set D of X

With the following definition we describe a large class of subsets D of W for which the basic inequality, proved in the next chapters, holds. The norming sets of the H.I. spaces we deal with are of such a form.

DEFINITION 3.8. A subset D of $W = W\left[G, \left(\mathcal{S}_{n_j}, \frac{1}{m_j}\right)_j\right]$ is said to be *tree complete* if:

(i) The set D is symmetric and $G \subset D$.

(ii) For every $f \in D$ there exists a tree $T_f = (f_a)_{a \in \mathcal{A}}$ of f such that $f_a \in D$ for every $a \in \mathcal{A}$.

(iii) For every $f \in D$ and interval E of the natural numbers we have that $Ef \in D$.

(iv) For infinitely many j's we have that D is j closed, that is if $(f_1, f_2, \dots, f_d)$ is an $\mathcal{S}_{n_j}$ admissible sequence of D then $\frac{1}{m_j}(f_1 + f_2 + \cdots + f_d) \in D$.

DEFINITION 3.9. Let D be a tree complete subset of W. We consider the norm $\| \ \| = \| \ \|_D$ on c_{00} defined by $\|x\| = \sup\{f(x) : f \in D\}$. The space $(X, \| \ \|)$, is the completion of c_{00} with the above defined norm. We denote $X = T\left[G, \left(\mathcal{S}_{n_j}, \frac{1}{m_j}\right)_j, D\right]$.

REMARK 3.10. Let $(x_k)_{k=1}^n$ be a block sequence in X such that $\|x_k\| \geq a > 0$ for each k, and $j \in \mathbb{N}$ such that D is j closed (Definition 3.8 (iv)) and the family $(x_k)_{k=1}^n$ is $\mathcal{S}_{n_j}$ admissible. Then for every convex combination $\sum_{k=1}^{n} a_k x_k$ we have that:

$$\|\sum_{k \in F} a_k x_k\| \geq \frac{a}{m_j}.$$

Since $W = \operatorname{conv}_{\mathbb{Q}}(A)$ and D is tree complete we have that $\|x\| = \sup\{f(x) : f \in A \cap D\}$ for every $x \in c_{00}$.

In the next proposition we list some properties of the space X that follow immediately from the definition of the norm of the space and the properties of the tree complete set D.

PROPOSITION 3.11. (i) The sequence $(e_n)_{n\in\mathbb{N}}$ is a normalized bimonotone and boundedly complete Schauder basis of X and for $x \in c_{00}$ $\|x\|_G \leq \|x\|_D$.
(ii) The space $X_* = \overline{\text{span}}\{e_n^* : n \in \mathbb{N}\}$ is a predual of X.
(iii) $B_{X^*} = \overline{\text{conv}(A \cap D)}^{w^*} = \overline{\text{conv}(A \cap D)}^{p}$ where A is the set defined in Definition 3.3.
(iv) The space X with the basis $(e_n)_{n\in\mathbb{N}}$ is an asymptotic ℓ^1 space.

NOTATION 3.12. In the sequel whenever we write m_j or n_j we shall refer to the following recursively defined sequences of natural numbers

$$m_1 = 2 \qquad n_1 = 1$$
$$m_{j+1} = {m_j}^{m_j} \quad n_{j+1} = 2^{2m_{j+1}} n_j.$$

Let's observe for later use that since the numbers m_j are powers of 2 the numbers $\log_2(m_j)$ are integers. The double sequence $(m_j, n_j)_j$ is one of the important ingredients in the definition of our spaces. However, it is worthwhile noting there are many others serving the same purpose. We recommend the interested reader the paper by G. Androulakis and E. Odell [**4**] where sufficient conditions for such sequences are given.

DEFINITION 3.13 (The auxiliary space $\boldsymbol{T_{ad}}$). Let $G_0 = \{\pm e_q : \quad q \in \mathbb{N}\}$. The auxiliary space T_{ad} is the completion of c_{00} under the norming set $D_0 = W[G_0, (\mathcal{S}_{n_j+1}, \frac{1}{m_j})_j]$.

If $f = \sum a_i e_i \in D_0$ then for every choice of $\varepsilon_i \in \{-1, 1\}$ also $\sum \varepsilon_i a_i e_i \in D_0$. So the basis $(e_i)_{i\in\mathbb{N}}$ of T_{ad} is 1-unconditional.

In the inductive definition of the set $W = \bigcup_{n=0}^{\infty} W_n$, each $W_n = \text{conv}_{\mathbb{Q}}(A_n)$. As we have implied in the beginning of this chapter this seems to be necessary when studying X^* and $\mathcal{L}(X)$ for X H.I. space constructed with these methods. This assumption creates some additional difficulties in estimating the norm of certain vectors. The lemma following the next definition will help us to overcome these difficulties and roughly speaking says that in the space T_{ad} there exist a subset of W not containing rational convex combinations and defining isometrically the norm of the space.

DEFINITION 3.14. We set $B = \bigcup_{n=0}^{\infty} B_n$ where the sequence $(B_n)_{n\in\mathbb{N}}$ is recursively defined as follows:

$$\begin{aligned} B_0 &= \{\pm e_q : \quad q \in \mathbb{N}\} \\ B_{n+1} &= \Big\{\frac{1}{m_j}(f_1 + f_2 + \cdots + f_d) : \quad f_i \in B_n \text{ for } i = 1, 2, \ldots, d \\ &\qquad \text{and the sequence } (f_1, f_2, \ldots, f_d) \text{ is } \mathcal{S}_{n_j+1} \text{ admissible, } j \in \mathbb{N}\Big\}. \end{aligned}$$

The set B is tree complete and every $f \in B$ has a tree $T_f = (f_a)_{a\in\mathcal{A}}$ such that each f_a is not of type II.

LEMMA 3.15. For all $j = 1, 2, \dots$ we have that

$$\operatorname{conv}_{\mathbb{Q}}\{f \in A : w(f) = m_j\} = \operatorname{conv}_{\mathbb{Q}}\{f \in B : w(f) = m_j\}.$$

This implies in particular that $W = \operatorname{conv}_{\mathbb{Q}}(B)$.

Proof. It is enough to prove that for all $j \in \mathbb{N}$ and $n = 1, 2, \dots$ we have that

(i) $\operatorname{conv}_{\mathbb{Q}}\{f \in A_n : w(f) = m_j\} = \operatorname{conv}_{\mathbb{Q}}\{f \in B_n : w(f) = m_j\}$.
(ii) $\{f \in B_n : w(f) = m_j\} \subset \{f \in A_n : w(f) = m_j\}$.

We prove (i) and (ii) by induction on n. The proof of the first inductive step is similar to the proof of the general inductive step; we proceed to the general inductive step. Suppose that (i) and (ii) hold for some n and let $f \in A_{n+1}$ with $w(f) = m_j$. Then $f = \frac{1}{m_j}\sum_{i=1}^{d} f_i$ where $(f_1, \dots, f_d)$ is an $\mathcal{S}_{n_j+1}$ admissible family of successive functionals of W_n. Each f_i belongs to $\operatorname{conv}_{\mathbb{Q}}(A_n)$ thus by our inductive assumption $f_i \in \operatorname{conv}_{\mathbb{Q}}(B_n)$. We can write $f_i = \frac{1}{\nu}\sum_{q=1}^{\nu} f_i^q$ with $f_i^q \in B_n$ and $\operatorname{supp} f_i^q \subset \operatorname{ran} f_i$ where the number ν is common for all i. We set $f^q = \frac{1}{m_j}\sum_{i=1}^{d} f_i^q$ for $q = 1, 2, \dots, \nu$. Clearly $f^q \in B_{n+1}$ with $w(f) = m_j$ hence the functional $f = \frac{1}{\nu}\sum_{q=1}^{\nu} f^q$ belongs to $\operatorname{conv}_{\mathbb{Q}}\{f \in B_{n+1} : w(f) = m_j\}$. Thus $\{f \in A_{n+1} : w(f) = m_j\} \subset \operatorname{conv}_{\mathbb{Q}}\{f \in B_{n+1} : w(f) = m_j\}$ and taking rational convex hulls we have (i) for $n+1$.

Let $f \in B_{n+1}$ with $w(f) = m_j$. Then $f = \frac{1}{m_j}\sum_{i=1}^{d} f_i$ for an $\mathcal{S}_{n_j+1}$ admissible sequence $(f_1, \dots, f_d)$ of successive functionals in B_n. By our inductive assumptions each f_i belongs to A_n and thus to W_n. The definition of A_{n+1} yields that this set contains the functional f, and since clearly $w(f) = m_j$, the proof of (ii) for $n+1$ is finished. □

LEMMA 3.16. Let $f \in B$.

(i) For $j = 2, 3, \dots$ we have $\{k \in \mathbb{N} : \ |f(e_k)| > \frac{1}{m_j}\} \in \mathcal{S}_{(\log_2(m_j)-1)(n_{j-1}+1)}$.
(ii) If $j = 2, 3, \dots$ is such that there exist a tree $T_f = (f_a)_{a\in\mathcal{A}}$ of f with $w(f) \neq m_j$ for each $a \in \mathcal{A}$ then
$\{k \in \mathbb{N} : \ |f(e_k)| > \frac{1}{m_j^2}\} \in \mathcal{S}_{(2\log_2(m_j)-1)(n_{j-1}+1)}$.

Proof. (i) Without loss of generality, we may assume that $f(e_k) > \frac{1}{m_j}$ for each $k \in \operatorname{supp} f$. We consider a tree $T_f = (f_a)_{a\in\mathcal{A}}$ of f such that each f_a is not of type II and let n be its height. Therefore the tree T_f has a branch $\Gamma = (f_{a_i})_{i=0}^{n}$ such that $f_{a_0} = f$, $f_{a_n} = e_q$ for some $q \in \mathbb{N}$ and the functionals f_{a_i}, $i = 0, 1, \dots, n$ are pairwise different. So for each $i = 0, 1, \dots, n-1$ f_{a_i} is of the form

$$f_{a_i} = \frac{1}{m_{j_i}} \sum_{\beta \in S_{f_{a_i}}} f_\beta$$

where $\{f_\beta : \beta \in S_{f_{\gamma_i}}\}$ is an $\mathcal{S}_{n_{j_i}+1}$ admissible family of W. Since $m_{j_i} \geq 2$ for all i, for each $k \in \operatorname{supp} f$ we have that $\frac{1}{m_j} < f(e_k) = \prod_{i=0}^{n-1} \frac{1}{m_{j_i}} \leq (\frac{1}{2})^n$ hence

$$n \leq \log_2(m_j) - 1 \tag{5}$$

Observe also that our assumption $f(e_k) > \frac{1}{m_j}$ for each $k \in \operatorname{supp} f$ yields that $w(f_a) \leq m_{j-1}$ for all $a \in \mathcal{A}$ and thus that the family $\{f_\beta : \beta \in S_a\}$ is $\mathcal{S}_{n_{j-1}+1}$ admissible.

Claim For each $i = 0, 1, \ldots, n$ the family $\{f_a : |a| = i\}$ is $\mathcal{S}_{i(n_{j-1}+1)}$ admissible.

Proof. We proceed by induction. For $i = 0$ the conclusion is obvious. Suppose that for some $i < n$ the family $\{f_a : |a| = i\}$ is $\mathcal{S}_{i(n_{j-1}+1)}$ admissible. Since for each such a the family $\{f_\beta : \beta \in S_a\}$ is $\mathcal{S}_{n_{j-1}+1}$ admissible we conclude that the family $\{f_a : |a| = i+1\}$ is $\mathcal{S}_{i(n_{j-1}+1)} * \mathcal{S}_{n_{j-1}+1} = \mathcal{S}_{(i+1)(n_{j-1}+1)}$ admissible. □

Hence $\{e_q : q \in \operatorname{supp} f\}$ is $\mathcal{S}_{n(n_{j-1}+1)}$ admissible therefore by (5) $\operatorname{supp} f \in \mathcal{S}_{(\log_2(m_j)-1)(n_{j-1}+1)}$.

(ii) The proof follows the same lines as that of (i). Let $T_f = (f_a)_{a \in \mathcal{A}}$ be a tree of f such that each f_a not of type II and moreover $w(f_a) \neq m_j$ for all $a \in \mathcal{A}$. We may assume that $f(e_k) > \frac{1}{m_j^2}$ for every $k \in \operatorname{supp} f$. The height of T_f is at most $2\log_2(m_j) - 1$. Since $m_{j+1} \geq m_j^2$ we get that $w(f_a) \leq m_{j-1}$ for all $a \in \mathcal{A}$. Hence $\operatorname{supp} f \in \mathcal{S}_{(2\log_2(m_j)-1)(n_{j-1}+1)}$. □

Definition 3.17 (basic special convex combinations). Let $\varepsilon > 0$ and $j \in \mathbb{N}$, $j > 1$. A convex combination $\sum_{k \in F} a_k e_k$ of the basis $(e_k)_{k \in \mathbb{N}}$ is said to be an (ε, j) *basic special convex combination* ((ε, j) B.S.C.C.) if

(1) $F \in \mathcal{S}_{n_j}$
(2) For every $P \in \mathcal{S}_{2\log_2(m_j)(n_{j-1}+1)}$ we have that $\sum_{k \in P} a_k < \varepsilon$.

Remark 3.18. If $\varepsilon > 0$, $j \in \mathbb{N}$ with $j > 1$ and $L \in [\mathbb{N}]$ with $\min L \geq \frac{3}{\varepsilon}$ then for each $q \in \mathbb{N}$ with $2\log_2(m_j)(n_{j-1}+1) < q \leq n_j$ the convex combination of the basis $a(q, L)$ is an (ε, j) B.S.C.C. Indeed

$$\operatorname{supp} a(q, L) \in \mathcal{S}_q \subset \mathcal{S}_{n_j}$$

while for every $P \in \mathcal{S}_{2\log_2(m_j)(n_{j-1}+1)}$ Proposition 2.3 yields that

$$\sum_{k \in P} a(q, L)(k) < \frac{3}{\min L} \leq \varepsilon.$$

Part (i) of the next proposition is contained in [**6**]. For the sake of completeness we include the proof here.

Proposition 3.19 (Estimates on the basis of T_{ad}). Let $\varepsilon > 0$, $j \in \mathbb{N}$ with $j > 1$ such that $\varepsilon \leq \frac{1}{m_j}$ and $x = \sum_{k \in F} a_k e_k$ an (ε, j) B.S.C.C. Then:

(i) For every $f \in B$

$$|f(x)| \leq \begin{cases} \varepsilon & \text{if } f = \pm e_q^* \\ \dfrac{1}{m_i} & \text{if } w(f) = m_i \ , i \geq j \\ \dfrac{2}{m_i m_j} & \text{if } w(f) = m_i \ , i < j \end{cases}$$

and thus $|f(x)| \leq \dfrac{1}{m_j}$.

(ii) If the functional $f \in B$ has a tree $T_f = (f_a)_{a \in \mathcal{A}}$ with $w(f_a) \neq m_j$ for all $a \in \mathcal{A}$ and $\varepsilon \leq \dfrac{1}{m_j^2}$ then $|f(x)| \leq \dfrac{1}{m_j^2}$.

Proof. We proceed to the proof of (i). We may assume that $\operatorname{supp} f \subset F$ and $f(e_k) \geq 0$ for each $k \in \mathbb{N}$.
If $f = e_q^*$ for $q \in F$ then $f(x) = a_q < \varepsilon$ since x is an (ε, j) B.S.C.C. and $\{q\} \in \mathcal{S}_0 \subset \mathcal{S}_{2\log_2(m_j)(n_{j-1}+1)}$.
If $w(f) = m_i$ for $i \geq j$ then we get $f(x) \leq \dfrac{1}{m_i}$ since $\|f\|_\infty \leq \dfrac{1}{m_i}$ and $\sum\limits_{k \in F} a_k = 1$.

Let now $f \in B$ with $w(f) = m_i$ for $i < j$. Then $f = \dfrac{1}{m_i}(f_1 + f_2 + \cdots + f_d)$ for some $\mathcal{S}_{n_i+1}$ admissible sequence $(f_1, f_2, \ldots, f_d)$ in B. For $t = 1, 2, \ldots, d$ we set

$$D_t = \{q \in \mathbb{N} : \ f_t(e_q) > \frac{1}{m_j}\} \quad \text{and} \quad D = \bigcup_{t=1}^{d} D_t.$$

From Lemma (3.16) we get that $D_t \in \mathcal{S}_{(\log_2(m_j)-1)(n_{j-1}+1)}$ for each $t = 1, 2, \ldots, d$. Hence, since the sequence $(f_1, f_2, \ldots, f_d)$ is $\mathcal{S}_{n_i+1}$ admissible and thus (since $i \leq j-1$) is $\mathcal{S}_{n_{j-1}+1}$ admissible and $D_t \subset \operatorname{supp} f_t$ for each $t = 1, 2, \ldots, d$ we conclude that the sequence $(D_1 < D_2 < \cdots < D_d)$ is $\mathcal{S}_{n_{j-1}+1}$ admissible. Thus

$$D \in \mathcal{S}_{n_{j-1}+1} * \mathcal{S}_{(\log_2(m_j)-1)(n_{j-1}+1)} = \mathcal{S}_{\log_2(m_j)(n_{j-1}+1)}. \tag{6}$$

Since $x = \sum\limits_{k \in F} a_k e_k$ is an (ε, j) B.S.C.C., (6) implies that $\sum\limits_{q \in D} a_q < \varepsilon$. Therefore

$$\begin{aligned} f(x) &= \frac{1}{m_i} \sum_{t=1}^{d} f_t(\sum_{k \in F} a_k e_k) = \frac{1}{m_i}(\sum_{t=1}^{d} f_t|_D(\sum_{k \in F} a_k e_k) + \sum_{t=1}^{d} f_t|_{\mathbb{N}\setminus D}(\sum_{k \in F} a_k e_k)) \\ &\leq \frac{1}{m_i}(\sum_{k \in D} a_k + \frac{1}{m_j}) < \frac{1}{m_i}(\varepsilon + \frac{1}{m_j}) \\ &\leq \frac{1}{m_i}(\frac{1}{m_j} + \frac{1}{m_j}) = \frac{2}{m_i m_j} \end{aligned}$$

The proof of (ii) follows the same lines so we omit it. □

CHAPTER 4

The basic inequality

All the proofs yielding that a Banach space is H.I., with the exception of Gowers dichotomy [**27**], requires at some stage the computation of the norm of certain vectors. In the case of asymptotic ℓ^1 spaces [**6**], the following was used. For a bounded block sequence $(x_k)_{k\in\mathbb{N}}$ satisfying some additional properties and any $f \in D$, using the tree T_f of the analysis of f, a functional g acting on the space T_{ad} was constructed such that $|f(x_k)| \leq C \cdot g(e_{t_k})$ for all k. This permits to reduce the computation of $f(\sum a_k x_k)$ to $g(\sum a_k e_{t_k})$ and get the desired results. The present chapter contains a similar procedure in the framework of $X = T[G, (\mathcal{S}_{n_j}, \frac{1}{m_j})_j, D]$ and we have attempted to formulate the results in the most general setting. This will enable us to use the basic inequality in several different stages.

DEFINITION 4.1. Let $(x_k)_{k\in\mathbb{N}}$ be a block sequence and $\varepsilon > 0$. We set $t_k = \min\operatorname{supp} x_k$ for each $k = 1, 2, \ldots$ The sequence $(x_k)_{k\in\mathbb{N}}$ is said to be (G, ε) Schreier if for every $g \in G$ the subset of the natural numbers

$$\{t_k : \ k \in \mathbb{N} \text{ such that } |g(x_k)| > \varepsilon\}$$

belongs to the second Schreier family $\mathcal{S}_2$.

NOTATION 4.2. Throughout this chapter for a ground set G and D a tree complete subset of the set $W = W[G, (\mathcal{S}_{n_j}, \frac{1}{m_j})_j]$ we denote by X the space defined by the norming set D. Furthermore we denote by A the subset of $W[G, (\mathcal{S}_{n_j}, \frac{1}{m_j})_j]$ as it is defined in Definition 3.3 while $W' = W[G_0, (\mathcal{S}_{n_j+1}, \frac{1}{m_j})_j]$ and A' are the corresponding sets for the space T_{ad} (Definition 3.13) and B is the set described in Definition 3.14.

PROPOSITION 4.3 (The basic inequality). Let $(x_k)_{k\in\mathbb{N}}$ be a block sequence in X, $(j_k)_{k\in\mathbb{N}}$ a strictly increasing sequence of natural numbers, $(\lambda_k)_{k\in\mathbb{N}} \in c_{00}$, $C \geq 1$ and $\varepsilon > 0$ such that the following are fulfilled:

(a) The sequence $(x_k)_{k\in\mathbb{N}}$ is (G, ε) Schreier with $\|x_k\| \leq C$ for all k and $\min\operatorname{supp} x_1 \geq 7$.
(b) $(\max\operatorname{supp} x_k)\frac{1}{\varepsilon} \leq m_{j_{k+1}}$ for each $k = 1, 2, \ldots$
(c) For each $k = 1, 2, \ldots$ and $h \in D$ with $w(h) < m_{j_k}$ we have that $|h(x_k)| \leq \frac{C}{w(h)}$.

Then for every $f \in A \cap D$ we can find g_1, such that either $g_1 = h_1$ or $g_1 = e_t^* + h_1$ with $t \notin \operatorname{supp} h_1$ where $h_1 \in \operatorname{conv}_{\mathbb{Q}}\{h \in B : \ w(h) = w(f)\}$ and $g_2 \in c_{00}$ with

$\|g_2\|_\infty \leq \varepsilon$ such that

$$|f(\sum \lambda_k x_k)| \leq C(2g_1 + g_2)(\sum |\lambda_k| e_{t_k}) \tag{7}$$

where $t_k = \min \operatorname{supp} x_k$ for each $k = 1, 2, \ldots$.

If we additionally assume that for some $j_0 \in \mathbb{N}$ we have that

(d) For every $h \in A \cap D$ with $w(h) = m_{j_0}$ and every interval E of the natural numbers

$$|h(\sum_{k\in E} \lambda_k x_k)| \leq C(2 \max_{k\in E} |\lambda_k| + \varepsilon \sum_{k\in E} |\lambda_k|) \tag{8}$$

then h_1 may be selected satisfying the following property: $h_1 = \sum r_i \tilde{h}_i$ with $r_i \in \mathbb{Q}^+$, $\sum r_i = 1$ and for each i the functional $\tilde{h}_i$ belongs to B and has a tree $T_{\tilde{h}_i} = (f_a^i)_{a\in\mathcal{C}_i}$ with $w(f_a^i) \neq m_{j_0}$ for all $a \in \mathcal{C}_i$.

REMARK 4.4. Our intention is to apply the above inequality in order to obtain upper estimates for special convex combinations of rapidly increasing sequences. Observe that the above proposition reduces this problem to estimates of the functionals g_1, g_2 on a corresponding basic S.C.C. in T_{ad}. Since ε can be chosen arbitrarily small, the g_2 estimate becomes negligible while the g_1 estimate follows from Proposition 3.19.

The proof in the general case (assuming only $(a), (b), (c)$) and in the special case (where additionally (d) is assumed) is the same. We give the proof only in the special case. The proof in the general case arises by omitting any reference to distinguishing cases whether a functional has weight m_{j_0} or not.

For the remainder we assume that there exists $j_0 \in \mathbb{N}$ such that condition (d) in the statement of Proposition 4.3 is satisfied. We also fix an $f \in A \cap D$ and a tree (Definition 3.5) $T_f = (f_a)_{a\in\mathcal{A}}$ of f with each $f_a \in D$. Before passing to the proof we adopt some useful notation and state two lemmas.

DEFINITION 4.5. Let $(x_k)_{k\in\mathbb{N}}$ be a block sequence in X, $j_0 \in \mathbb{N}$, $f \in A \cap D$ and $T_f = (f_a)_{a\in\mathcal{A}}$ as before. For each $k \in \mathbb{N}$ we define the set $\mathcal{A}_k$ as follows:

$$\begin{aligned}\mathcal{A}_k = \Big\{ & a \in \mathcal{A} \text{ such that } f_a \text{ is not of type } II \text{ and} \\ & (i)\ \operatorname{ran} f_a \cap \operatorname{ran} x_k \neq \emptyset \\ & (ii)\ \forall\, \gamma < a \text{ if } f_\gamma \text{ is of type } I \text{ then } w(f_\gamma) \neq m_{j_0} \\ & (iii)\ \forall\, \beta \leq a \ \text{ if } \beta \in S_\gamma \text{ and } f_\gamma \text{ is of type } I \\ & \text{then } \operatorname{ran} f_\beta \cap \operatorname{ran} x_k = \operatorname{ran} f_\gamma \cap \operatorname{ran} x_k \\ & (iv)\ \text{if } w(f_a) \neq m_{j_0} \text{ then for all } \beta \in S_a \\ & \operatorname{ran} f_\beta \cap \operatorname{ran} x_k \subsetneqq \operatorname{ran} f_a \cap \operatorname{ran} x_k \Big\}\end{aligned}$$

The next lemma describes the properties of the set $\mathcal{A}_k$.

LEMMA 4.6. For every $k \in \mathbb{N}$ we have the following:

(i) If $a \in \mathcal{A}$ and f_a is of type II then $a \notin \mathcal{A}_k$.
(Hence $\mathcal{A}_k \subset \{a \in \mathcal{A} : f_a \text{ is of type } I \text{ or } f_a \in G\}$.)
(ii) If $a \in \mathcal{A}_k$, then for every $\beta < a$ if f_β is of type I then $w(f_\beta) \neq m_{j_0}$.
(iii) If $\mathcal{A}_k$ is not a singleton then its members are incomparable members of the tree $\mathcal{A}$. Moreover if a_1, a_2 are two different elements of $\mathcal{A}_k$ and β

is the (necessarily uniquely determined) maximal element of $\mathcal{A}$ satisfying $\beta < a_1$ and $\beta < a_2$ then f_β is of type II.

(iv) If $a \in \mathcal{A}$ is such that $\operatorname{supp} f_a \cap \operatorname{ran} x_k \neq \emptyset$ and $\gamma \notin \mathcal{A}_k$ for all $\gamma < a$ then there exists $\beta \in \mathcal{A}_k$ with $a \leq \beta$. In particular if $\operatorname{supp} f \cap \operatorname{ran} x_k \neq \emptyset$ then $\mathcal{A}_k \neq \emptyset$.

Proof. Properties (i) and (ii) follow immediately from the definition of the set $\mathcal{A}_k$. Let's prove property (iii). Suppose that the set $\mathcal{A}_k$ contains two comparable members of the tree $\mathcal{A}$; let $a_1, a_2 \in \mathcal{A}_k$ with $a_1 < a_2$. Since a_1 is not a maximal element of $\mathcal{A}$, we have $f_{a_1} \notin G$ and from property (i) we get that f_{a_1} is of type I. Condition (ii) of Definition 4.5 for a_2 yields that $w(f_{a_1}) \neq m_{j_0}$. We denote by β the immediate successor of a_1 in the path connecting a_1 with a_2. Condition (iii) of Definition 4.5 for a_2, and the facts that $\beta \leq a_2$, $\beta \in S_{a_1}$ and f_{a_1} is of type I, yield that $\operatorname{ran} f_\beta \cap \operatorname{ran} x_k = \operatorname{ran} f_{a_1} \cap \operatorname{ran} x_k$. On the other hand, since f_{a_1} is of type I with $w(f_{a_1}) \neq m_{j_0}$, from condition (iv) of Definition 4.5 for a_1 we get that $\operatorname{ran} f_\beta \cap \operatorname{ran} x_k \subsetneqq \operatorname{ran} f_{a_1} \cap \operatorname{ran} x_k$. This contradiction shows that the members of $\mathcal{A}_k$ are incomparable nodes of the tree $\mathcal{A}$. Let now $a_1, a_2 \in \mathcal{A}_k$ be two different elements of $\mathcal{A}_k$ and let β be the maximal element of $\mathcal{A}$ satisfying $\beta < a_1$ and $\beta < a_2$. We denote by β_1 (resp. β_2) the immediate successor of β in the path connecting β with a_1 (resp. a_2). If f_β is of type I then condition (iii) in the definition of $\mathcal{A}_k$ for a_1 and a_2 yield that $\operatorname{ran} f_{\beta_1} \cap \operatorname{ran} x_k = \operatorname{ran} f_\beta \cap \operatorname{ran} x_k$ and $\operatorname{ran} f_{\beta_2} \cap \operatorname{ran} x_k = \operatorname{ran} f_\beta \cap \operatorname{ran} x_k$. But since f_{β_1}, f_{β_2} are successive and hence disjointly supported, we get that $\operatorname{ran} f_\beta \cap \operatorname{ran} x_k = \emptyset$, and hence $\operatorname{ran} f_{a_1} \cap \operatorname{ran} x_k = \emptyset$ contradicting condition (i) of Definition 4.5 for a_1. Therefore f_β is of type II.

We next prove property (iv) using backward induction. The proof of the first inductive step (for $a \in \mathcal{A}$ which are maximal) and in the general inductive step is the same. Let $a \in \mathcal{A}$ such that property (iv) is true for all $\gamma \in \mathcal{A}$ with $\gamma > a$. We assume that $\operatorname{supp} f_a \cap \operatorname{ran} x_k \neq \emptyset$ and $\gamma \notin \mathcal{A}_k$ for all $\gamma < a$. Then for every $\gamma < a$ with f_γ of type I the following two conditions are satisfied. First $w(f_\gamma) \neq m_{j_0}$, and second, if $\beta \in S_\gamma$ with $\beta \leq a$ then $\operatorname{ran} f_\beta \cap \operatorname{ran} x_k = \operatorname{ran} f_\gamma \cap \operatorname{ran} x_k$ (indeed if this is not the case and γ is the minimal $\gamma < a$ with f_γ of type I failing to satisfy one of these two conditions, then $\gamma \in \mathcal{A}_k$, a contradiction). If f_a is of type II, then $a \notin \mathcal{A}_k$ and there exists $\gamma \in S_a$ such that $\operatorname{supp} f_\gamma \cap \operatorname{ran} x_k \neq \emptyset$. The inductive assumptions are satisfied for γ, thus there exists $\beta \in \mathcal{A}_k$ with $a < \gamma \leq \beta$. If f_a is of type I, we distinguish two cases. If $w(f_a) = m_{j_0}$ then $a \in \mathcal{A}_k$. If $w(f_a) \neq m_{j_0}$ then there are two subcases. Either $\operatorname{ran} f_\beta \cap \operatorname{ran} x_k \subsetneqq \operatorname{ran} f_a \cap \operatorname{ran} x_k$ for all $\beta \in S_a$, and hence $a \in \mathcal{A}_k$, or there exists $\gamma \in S_a$ such that $\operatorname{ran} f_\gamma \cap \operatorname{ran} x_k = \operatorname{ran} f_a \cap \operatorname{ran} x_k$. In the later case, from the fact that $\operatorname{supp} f_a \cap \operatorname{ran} x_k \neq \emptyset$ we get that $\operatorname{supp} f_\gamma \cap \operatorname{ran} x_k \neq \emptyset$. The inductive assumptions are fulfilled for γ and therefore there exists $\beta \in \mathcal{A}_k$ with $a < \gamma \leq \beta$. The proof of the lemma is complete. □

DEFINITION 4.7. For every $a \in \mathcal{A}$ we define $D_a = \bigcup_{\beta \geq a} \{k : \beta \in \mathcal{A}_k\}$.

LEMMA 4.8. According to the notation above we have the following:

(i) If $\operatorname{supp} f \cap \operatorname{ran} x_k \neq \emptyset$ then $k \in D_0$ (remember that 0 denotes the unique root of $\mathcal{A}$ and $f = f_0$). Hence $f(\sum \lambda_k x_k) = f(\sum_{k \in D_0} \lambda_k x_k)$.

(ii) If f_a is of type I with $w(f_a) = m_{j_0}$ then D_a is an interval of $\mathbb{N}$.

(iii) If f_a is of type I with $w(f_a) \neq m_{j_0}$ then

$$\left\{\{k\} : k \in D_a \setminus \bigcup_{\beta \in S_a} D_\beta\right\} \cup \{D_\beta : \beta \in S_a\}$$

is a family of successive subsets of $\mathbb{N}$. Moreover for every $k \in D_a \setminus \bigcup_{\beta \in S_a} D_\beta$ (i.e. for k such that $a \in \mathcal{A}_k$) such that $\operatorname{supp} f_a \cap \operatorname{ran} x_k \neq \emptyset$ there exists a $\beta \in S_a$ such that either $\min\operatorname{supp} x_k \leq \max\operatorname{supp} f_\beta < \max\operatorname{supp} x_k$ or $\min\operatorname{supp} x_k < \min\operatorname{supp} f_\beta \leq \max\operatorname{supp} x_k$.

(iv) If f_a is of type II, $\beta \in S_a$ and $k \in D_a \setminus D_\beta$ then $\operatorname{supp} f_\beta \cap \operatorname{ran} x_k = \emptyset$ and hence $f_\beta(x_k) = 0$.

Proof. If $\operatorname{supp} f \cap \operatorname{ran} x_k \neq \emptyset$ then, from Lemma 4.6(iv), we get that $\mathcal{A}_k \neq \emptyset$ and hence $k \in D_0$. Thus assertion (i) holds.

To see (ii) let $a \in \mathcal{A}$ such that f_a is of type I with $w(f_a) = m_{j_0}$. Then, $\gamma \notin \mathcal{A}_k$ for all $\gamma > a$ and all k (Lemma 4.6(ii)), hence $D_a = \{k : a \in \mathcal{A}_k\}$. Let $k_1 < k < k_2$ with $k_1, k_2 \in D_a$, i.e. $a \in \mathcal{A}_{k_1}$ and $a \in \mathcal{A}_{k_2}$. We shall show that $a \in \mathcal{A}_k$ (by showing that conditions $(i) - (iv)$ of Definition 4.5 are satisfied for a) and hence $k \in D_a$. Since $a \in \mathcal{A}_{k_1}$ for every $\gamma < a$ with f_γ of type I we have that $w(f_\gamma) \neq m_{j_0}$, hence condition (ii) is satisfied. We also have that $\operatorname{ran} f_a \cap \operatorname{ran} x_{k_i} \neq \emptyset$ for $i = 1, 2$ (since $a \in \mathcal{A}_{k_1} \cap \mathcal{A}_{k_2}$), thus $\min\operatorname{supp} f_a \leq \max\operatorname{supp} x_{k_1} < \min\operatorname{supp} x_k \leq \max\operatorname{supp} x_k < \min\operatorname{supp} x_{k_2} \leq \max\operatorname{supp} f_a$. This yields that $\operatorname{ran} x_k \subset \operatorname{ran} f_a$ and hence also conditions (i) and (iii) of Definition 4.5 are satisfied, therefore $a \in \mathcal{A}_k$ (condition (iv) is meaningless as $w(f_a) = m_{j_0}$).

Let's prove assertion (iii). Let f_a be of type I with $w(f_a) \neq m_{j_0}$ and observe that, as follows from Lemma 4.6(iii), the family in the statement consists of pairwise disjoint sets. To prove that these sets are successive, we have to show that if $k_1, k_2 \in D_\beta$ for some $\beta \in S_a$ and $k_1 < k < k_2$ then neither $a \in \mathcal{A}_k$, nor $k \in D_{\beta'}$ for any $\beta' \in S_a$, $\beta' \neq \beta$. Let $\beta \in S_a$, $k_1, k_2 \in D_\beta$ and $k_1 < k < k_2$. Then, similarly to the proof of (ii), we obtain that $\operatorname{ran} x_k \subset \operatorname{ran} f_\beta$. We do not have $a \in \mathcal{A}_k$ as condition (iv) of Definition 4.5 fails. Also for $\beta' \in S_a$, $\beta' \neq \beta$ we have $\operatorname{ran} f_{\beta'} \cap \operatorname{ran} x_k = \emptyset$, thus there is no $\gamma \geq \beta'$ with $\gamma \in \mathcal{A}_k$ since condition (iii) of Definition 4.5 fails, hence $k \notin D_{\beta'}$. We have thus proved the first part of (iii). To prove the second part, let k such that there is no $\beta \in S_a$ satisfying the conclusion claimed. Then either $\operatorname{supp} f_a \cap \operatorname{ran} x_k = \emptyset$ or there exists $\beta \in S_a$ such that $\operatorname{ran} x_k \subset \operatorname{ran} f_\beta$. In both cases it follows that $a \notin \mathcal{A}_k$.

We finally prove assertion (iv). Let f_a be of type II, $\beta \in S_a$ and $k \in D_a \setminus D_\beta$. By the definition of D_a there exists $\delta > a$ with $\delta \in \mathcal{A}_k$ (we can not have $a \in \mathcal{A}_k$ as f_a is of type II). Since $k \notin D_\beta$ we have $\beta \nleq \delta$. From the first part of Lemma 4.6(iii) $\gamma \notin \mathcal{A}_k$ for all $\gamma \leq a$, thus $\gamma \notin \mathcal{A}_k$ for all $\gamma < \beta$. Suppose now that $\operatorname{supp} f_\beta \cap \operatorname{ran} x_k \neq \emptyset$. Then Lemma 4.6(iv) yields that there exists $\zeta \geq \beta$ with $\zeta \in \mathcal{A}_k$. This contradicts to the second part of Lemma 4.6(iii), because $\delta, \zeta \in \mathcal{A}_k$, a is the maximal element of $\mathcal{A}$ satisfying $a < \delta$ and $a < \zeta$, and a is not of type I. Therefore $\operatorname{supp} f_\beta \cap \operatorname{ran} x_k = \emptyset$. □

Proof of the basic inequality. Let $f \in A \cap D$. We construct two families $(g_a^1)_{a \in \mathcal{A}}$ and $(g_a^2)_{a \in \mathcal{A}}$ such that the following are fulfilled.

(i) For every $a \in \mathcal{A}$ such that f_a is not of type II, $g_a^1 = h_a$ or $g_a^1 = e_{t_a}^* + h_a$ with $t_a \notin \operatorname{supp} h_a$, where $h_a \in \operatorname{conv}_{\mathbb{Q}}(B)$ and $g_a^2 \in c_{00}$ with $\|g_a^2\|_\infty \leq \varepsilon$.

(ii) For every $a \in \mathcal{A}$, $\operatorname{supp} g_a^1 \subset \{t_k : k \in D_a\}$ and $\operatorname{supp} g_a^2 \subset \{t_k : k \in D_a\}$ and the functionals g_a^1, g_a^2 have nonnegative coordinates.

(iii) For $a \in \mathcal{A}$ with $f_a \in G$ and $D_a \neq \emptyset$ we have that $g_a^1 \in B$ with $w(g_a^1) = m_1 (= 2)$.

(iv) For f_a of type II with $f = \sum_{\beta \in S_a} r_\beta f_\beta$ (where $r_\beta \in \mathbb{Q}^+$ for every $\beta \in S_a$ and $\sum_{\beta \in S_a} r_\beta = 1$) we have $g_a^1 = \sum_{\beta \in S_a} r_\beta g_\beta^1$ and $g_a^2 = \sum_{\beta \in S_a} r_\beta g_\beta^2$.

(v) For f_a of type I with $w(f) = m_{j_0}$ we have $g_a^1 = e_{t_{k_a}}^*$ where $k_a \in D_a$ is such that $|\lambda_{k_a}| = \max_{k \in D_a} |\lambda_k|$ and $g_a^2 = \sum_{k \in D_a} \varepsilon e_{t_k}^*$.

(vi) For f_a of type I with $w(f) = m_j$ for $j \neq j_0$ we have $g_a^1 = h_a$ or $g_a^1 = e_{t_a}^* + h_a$ with $h_a \in \operatorname{conv}_{\mathbb{Q}}\{g \in B : w(g) = m_j\}$ and $t_a \notin \operatorname{supp} h_a$.

(vii) For every $a \in \mathcal{A}$ the following inequality holds:

$$|f_a(\sum_{k \in D_a} \lambda_k x_k)| \leq C(2g_a^1 + g_a^2)(\sum_{k \in D_a} |\lambda_k| e_{t_k}).$$

When the construction of $(g_a^1)_{a \in \mathcal{A}}$ and $(g_a^2)_{a \in \mathcal{A}}$ has been accomplished, then setting $g_1 = g_0^1$ and $g_2 = g_0^2$ (where 0 is the root of $\mathcal{A}$ and $f = f_0$), we observe that these are the desired functionals. To show that such $(g_a^1)_{a \in \mathcal{A}}$ and $(g_a^2)_{a \in \mathcal{A}}$ exist we use finite induction starting with $a \in \mathcal{A}$ which are maximal and in the general inductive step we assume that g_β^1, g_β^2 have been defined for all $\beta > a$ satisfying the inductive assumptions and we define g_a^1 and g_a^2.

$1 \stackrel{st}{=}$ inductive step

Let $a \in \mathcal{A}$ which is maximal. Then $f_a \in G$. If $D_a = \emptyset$ we define $g_a^1 = 0$ and $g_a^2 = 0$. If $D_a \neq \emptyset$ we set

$$E_a = \{k \in D_a : |f_a(x_k)| > \varepsilon\} \quad \text{and} \quad F_a = D_a \setminus E_a.$$

Since $(x_k)_{k \in \mathbb{N}}$ is (G, ε) Schreier we get that $\{t_k : k \in E_a\} \in \mathcal{S}_2$ and we define

$$g_a^1 = \frac{1}{2} \sum_{k \in E_a} e_{t_k}^* \quad \text{and} \quad g_a^2 = \sum_{k \in F_a} \varepsilon e_{t_k}^*$$

We recall that $m_1 = 2$ and $n_1 + 1 = 1 + 1 = 2$ so $g_a^1 \in B$. It remains to show the inequality (vii). For each k we have that $|f_a(x_k)| \leq \|x_k\| \leq C$. Hence

$$\begin{aligned} |f_a(\sum_{k \in D_a} \lambda_k x_k)| &\leq \sum_{k \in D_a} |\lambda_k| |f_a(x_k)| \leq \sum_{k \in E_a} |\lambda_k| C + \sum_{k \in F_a} |\lambda_k| \varepsilon \\ &\leq C(\sum_{k \in E_a} |\lambda_k| + \sum_{k \in F_a} |\lambda_k| \varepsilon) = C(2g_a^1 + g_a^2)(\sum_{k \in D_a} |\lambda_k| e_{t_k}). \end{aligned}$$

General inductive step

Let $a \in \mathcal{A}$ and suppose that g_γ^1 and g_γ^2 have been defined for every $\gamma > a$ satisfying the inductive assumptions. If $D_a = \emptyset$ we set $g_a^1 = 0$ and $g_a^2 = 0$. In the remainder of the proof we assume that $D_a \neq \emptyset$. We consider the following three cases:

$1 \stackrel{st}{=}$ case The functional f_a is of type II.

Let $f_a = \sum_{\beta \in S_a} r_\beta f_\beta$ where $r_\beta \in \mathbb{Q}^+$ are such that $\sum_{\beta \in S_a} r_\beta = 1$. In this case, we have that $D_a = \bigcup_{\beta \in S_a} D_\beta$. We define

$$g_a^1 = \sum_{\beta \in S_a} r_\beta g_\beta^1 \quad \text{and} \quad g_a^2 = \sum_{\beta \in S_a} r_\beta g_\beta^2.$$

We show inequality (vii).

$$\begin{aligned}
|f_a(\sum_{k\in D_a}\lambda_k x_k)| &\leq \sum_{\beta\in S_a} r_\beta |f_\beta(\sum_{k\in D_a}\lambda_k x_k)| \\
&= \sum_{\beta\in S_a} r_\beta |f_\beta(\sum_{k\in D_\beta}\lambda_k x_k)| \quad \text{(by Lemma 4.8(iv))} \\
&\leq \sum_{\beta\in S_a} r_\beta C(2g_\beta^1+g_\beta^2)(\sum_{k\in D_\beta}|\lambda_k| e_{t_k}) \quad \text{(by ind. ass. (vii))} \\
&= \sum_{\beta\in S_a} r_\beta C(2g_\beta^1+g_\beta^2)(\sum_{k\in D_a}|\lambda_k| e_{t_k}) \quad \text{(by ind. ass. (ii))} \\
&= C(2\sum_{\beta\in S_a} r_\beta g_\beta^1 + \sum_{\beta\in S_a} r_\beta g_\beta^2)(\sum_{k\in D_a}|\lambda_k| e_{t_k}) \\
&= C(2g_a^1+g_a^2)(\sum_{k\in D_a}|\lambda_k| e_{t_k}).
\end{aligned}$$

$2 \stackrel{nd}{=}$ case The functional f_a is of type I with $w(f) = m_{j_0}$.
In this case D_a is an interval of the natural numbers (Lemma 4.8(ii)). Let $k_a \in D_a$ be such that $|\lambda_{k_a}| = \max_{k\in D_a} |\lambda_k|$. We define

$$g_a^1 = e_{t_{k_a}}^* \quad \text{and} \quad g_a^2 = \sum_{k\in D_a} \varepsilon e_{t_k}^*.$$

Then our assumption (d) in the statement of the proposition yields that

$$\begin{aligned}
|f_a(\sum_{k\in D_a}\lambda_k x_k)| &\leq C(2\max_{k\in D_a}|\lambda_k| + \varepsilon\sum_{k\in D_a}|\lambda_k|) \\
&= C(2e_{t_{k_a}}^* + \sum_{k\in D_a}\varepsilon e_{t_k}^*)(\sum_{k\in D_a}|\lambda_k| e_{t_k}) \\
&= C(2g_a^1+g_a^2)(\sum_{k\in D_a}|\lambda_k| e_{t_k}).
\end{aligned}$$

Hence inequality (vii) has been established.

$3 \stackrel{rd}{=}$ case The functional f_a is of type I with $w(f) = m_j$ for $j \neq j_0$.
Then $f_a = \frac{1}{m_j}\sum_{\beta\in S_a} f_\beta$ and the family $\{f_\beta : \beta \in S_a\}$ is S_{n_j} admissible. We set

$$\begin{aligned}
E_a &= \{k : a \in \mathcal{A}_k \text{ and } \operatorname{supp} f_a \cap \operatorname{ran} x_k \neq \emptyset\} \\
&\quad (= \{k \in D_a \setminus \bigcup_{\beta\in S_a} D_\beta : \operatorname{supp} f_a \cap \operatorname{ran} x_k \neq \emptyset\}).
\end{aligned}$$

We also consider the following partition of E_a.

$$E^2 = \{k \in E_a : m_{j_{k+1}} \leq m_j\} \quad \text{and} \quad E^1 = E_a \setminus E^2.$$

It follows from condition (b) in the statement of the proposition that for each $k \in E^2$ we have that

$$\begin{aligned}|f_a(x_k)| &\leq C\|f_a\|_\infty \#(\operatorname{supp} x_k) \leq C\frac{1}{m_j}\max\operatorname{supp} x_k\\ &\leq C\frac{1}{m_{j_{k+1}}}\max\operatorname{supp} x_k\\ &\leq C\varepsilon.\end{aligned}$$

We define

$$g_a^2 = \sum_{k\in E^2} \varepsilon e^*_{t_k} + \sum_{\beta\in S_a} g_\beta^2.$$

Observe that $\|g_a^2\|_\infty \leq \varepsilon$. Since for each $k \in E^2$ we have $g_a^2(e_{t_k}) = \varepsilon$, it follows that

(9) $$|f_a(x_k)| \leq C g_a^2(e_{t_k}) \text{ for all } k \in E^2$$

Let $E^1 = \{k_1 < k_2 < \cdots < k_l\}$. From the definition of E^1 we get that $m_j < m_{j_{k_2}} < \cdots < m_{j_{k_l}}$. Thus condition (c) in the statement of the proposition yields that

$$|f_a(x_{k_i})| \leq \frac{C}{m_j} \quad \text{for each } i = 2, \ldots, l.$$

We set

$$t_a = t_{k_1} \text{ and } g_a^1 = e^*_{t_a} + h_a \text{ where } h_a = \frac{1}{m_j}\Big(\sum_{i=2}^{l} e^*_{t_{k_i}} + \sum_{\beta\in S_a} g_\beta^1\Big)$$

(The term $e^*_{t_a}$ does not appear if $E^1 = \emptyset$.)
We observe that

(10) $$|f_a(x_k)| \leq C g_a^1(e_{t_k}) \quad \text{for all } k \in E^1.$$

(Indeed, for $k = k_1$ we have $|f_a(x_k)| \leq \|x_k\| \leq C = C g_a^1(e_{t_k})$ while for $k \in \{k_2, \ldots, k_l\}$ we have that $|f_a(x_k)| \leq \dfrac{C}{m_j} = C g_a^1(e_{t_k})$.)

We also have that

(11) $$\Big|f_a\Big(\sum_{k\in \bigcup_{\beta\in S_a} D_\beta} \lambda_k x_k\Big)\Big| \leq C(2g_a^1 + g_a^2)\Big(\sum_{k\in \bigcup_{\beta\in S_a} D_\beta} |\lambda_k| e_{t_k}\Big).$$

Indeed, from inductive assumption (vii) and the fact that the sets $(D_\beta)_{\beta\in S_a}$ are pairwise disjoint, we get that

$$\begin{aligned}\Big|f_a\Big(\sum_{k\in \bigcup_{\beta\in S_a} D_\beta} \lambda_k x_k\Big)\Big| &= \Big|\sum_{\beta\in S_a} f_a\Big(\sum_{k\in D_\beta}\lambda_k x_k\Big)\Big| = \Big|\sum_{\beta\in S_a}\frac{1}{m_j} f_\beta\Big(\sum_{k\in D_\beta}\lambda_k x_k\Big)\Big|\\ &\leq \sum_{\beta\in S_a}\frac{1}{m_j}\Big|f_\beta\Big(\sum_{k\in D_\beta}\lambda_k x_k\Big)\Big|\\ &\leq \sum_{\beta\in S_a}\frac{1}{m_j} C(2g_\beta^1 + g_\beta^2)\Big(\sum_{k\in D_\beta}|\lambda_k| e_{t_k}\Big)\\ &\leq C(2g_a^1 + g_a^2)\Big(\sum_{k\in \bigcup_{\beta\in S_a} D_\beta} |\lambda_k| e_{t_k}\Big).\end{aligned}$$

Inequalities (9), (10), (11) and the fact that for $k \in D_a \setminus \bigcup_{\beta \in S_a} D_\beta$ with $k \notin E_a$ we have that $f_a(x_k) = 0$ yield that

$$|f_a(\sum_{k \in D_a} \lambda_k x_k)| \leq C(2g_a^1 + g_a^2)(\sum_{k \in D_a} |\lambda_k| e_{t_k})$$

therefore inequality (vii) holds.

By the second part of Lemma 4.8(iii), for every $k \in E_a$ there exists an element of the set $N = \{\min \operatorname{supp} f_\beta, \max \operatorname{supp} f_\beta : \beta \in S_a\}$ belonging to $\operatorname{ran} x_k$. Since the family of functionals $\{f_\beta : \beta \in S_a\}$ is $\mathcal{S}_{n_j}$ admissible and using the fact that the family of sets $\mathcal{S}_{n_j}$ is hereditary and spreading we obtain that the set $\{t_k : k \in E_a \setminus \{\min E_a\}\}$ is the union of two sets belonging to S_{n_j}. Recalling that $E^1 = \{k_1 < k_2 < \cdots < k_l\}$, we get that $\{t_{k_2}, \ldots, t_{k_l}\}$ is the union of two S_{n_j} sets.

It remains to show that $h_a \in \operatorname{conv}_{\mathbb{Q}}\{g \in B : w(g) = m_j\}$. We first examine the case that for every $\beta \in S_a$ the functional f_β is not of type II. Then for every $\beta \in S_a$ one of the following holds:

(i) $f_\beta \in G$. In this case $g_\beta^1 \in B$ (by the first inductive step).

(ii) f_β is of type I with $w(f_\beta) = m_{j_0}$. In this case $g_\beta^1 = e_{t_{k_\beta}}^* \in B$.

(iii) f_β is of type I with $w(f_\beta) = m_j$ for $j \neq j_0$. In this case $g_\beta^1 = e_{t_\beta}^* + h_\beta$ where $h_\beta \in \operatorname{conv}_{\mathbb{Q}}(B)$ and $t_\beta \notin \operatorname{supp} h_\beta$. We set $E_\beta^1 = \{n \in \mathbb{N} : n < t_\beta\}$, $E_\beta^2 = \{n \in \mathbb{N} : n > t_\beta\}$ and $h_\beta^1 = E_\beta^1 h_\beta$, $h_\beta^2 = E_\beta^2 h_\beta$. The functionals h_β^1, $e_{t_\beta}^*$, h_β^2 are successive and belong to $W = \operatorname{conv}_{\mathbb{Q}}(B)$.

We set

$$\begin{aligned} T_a^1 &= \{\beta \in S_a : f_\beta \in G\} \\ T_a^2 &= \{\beta \in S_a : f_\beta \text{ of type } I \text{ and } w(f_\beta) = m_{j_0}\} \\ T_a^3 &= \{\beta \in S_a : f_\beta \text{ of type } I \text{ and } w(f_\beta) \neq m_{j_0}\}. \end{aligned}$$

We denote by $\beta_{\min}$ the element $\beta \in S_a$ with f_β minimal in the natural ordering. For every $\beta \in S_a \setminus \{\beta_{\min}\}$ and $k \in D_\beta$ we have that $\min \operatorname{supp} x_k \in \operatorname{ran} f_\beta$, i.e. $t_k \in \operatorname{ran} f_\beta$. From inductive assumption (ii) it follows that $\operatorname{supp} g_\beta \subset \operatorname{ran} f_\beta$ for every $\beta \in S_a \setminus \{\beta_{\min}\}$. This yields that each one of the sets $\{\min \operatorname{supp} g_\beta^1 : \beta \in T_a^1 \cup T_a^2\}$, $\{\min \operatorname{supp} h_\beta^1 : \beta \in T_a^3\}$, $\{t_\beta : \beta \in T_a^3\}$ and $\{\min \operatorname{supp} h_\beta^2 : \beta \in T_a^3\}$ is the union of a set belonging to S_{n_j} and a singleton. Since also the set $\{t_{k_2}, \ldots, t_{k_l}\}$ is the union of two S_{n_j} sets, we get that the union F of all the above sets can be written as the union of seven sets belonging to S_{n_j}. Since $\min F \geq 7$ (because $\min \operatorname{supp} x_1 \geq 7$) it follows that $F \in S_{n_j+1}$ (Proposition 2.1). The family of successive (Lemma 4.8(iii)) functionals of W,

$$\begin{aligned} &\{e_{t_{k_i}}^* : i = 2, \ldots, l\} \cup \{g_\beta^1 : \beta \in T_a^1\} \cup \{g_\beta^1 : \beta \in T_a^2\} \cup \\ &\cup \{h_\beta^1 : \beta \in T_a^3\} \cup \{e_{t_\beta}^* : \beta \in T_a^3\} \cup \{h_\beta^2 : \beta \in T_a^3\} \end{aligned}$$

is thus S_{n_j+1} admissible, and we get that $h_a \in A$ with $w(h_a) = m_j$ therefore

$$h_a \in \operatorname{conv}_{\mathbb{Q}}\{g \in B : w(g) = m_j\}.$$

We now examine the case that for some $\beta \in S_a$ the functional f_β is of type II. We set

$$B_2 = \{\beta \in S_a : f_\beta \text{ is of type } II\} \text{ and } B_1 = S_a \setminus B_2.$$

For every $\beta \in B_2$ the functional f_β is of the form

$$f_\beta = \sum_{\gamma \in S_\beta} r_\gamma f_\gamma \text{ with } r_\gamma \in \mathbb{Q}^+ \quad \text{for every } \gamma \in S_\beta \text{ and } \sum_{\gamma \in S_\beta} r_\gamma = 1$$

and none of the functionals $\{f_\gamma : \gamma \in S_\beta\}$ is of type II. We may choose a $k \in \mathbb{N}$ such that for every $\beta \in B_2$ the functional f_β can be written in the form

$$f_\beta = \frac{1}{k} \sum_{q=1}^{k} f_{\beta,q}$$

where each $f_{\beta,q}$ belongs to $\{f_\gamma : \gamma \in S_\beta\}$ for $q = 1, 2, \ldots, k$ (some of the functional $f_{\beta,q}$ may appear more than once in the above sum).

For each $q = 1, 2, \ldots, k$ we set

$$h_a^q = \frac{1}{m_j} \Big(\sum_{i=2}^{l} e_{k_i}^* + \sum_{\beta \in B_1} g_\beta^1 + \sum_{\beta \in B_2} g_{\beta,q}^1 \Big)$$

where $g_{\beta,q}^1 = g_\gamma^1$ for $\gamma \in S_\beta$ such that $f_{\beta,q} = f_\gamma$. According to what we have done just before we have

$$h_a^q \in \operatorname{conv}_{\mathbb{Q}}\{g \in B : w(g) = m_j\} \quad \text{for } q = 1, 2, \ldots, k.$$

Clearly $h_a = \frac{1}{k} \sum_{q=1}^{k} h_a^q$ hence we have that

$$h_a \in \operatorname{conv}_{\mathbb{Q}}\{g \in B : w(g) = m_j\}.$$

The proof of the general inductive step is complete and this finishes the whole proof of the proposition. □

CHAPTER 5

Special convex combinations in X

Special convex combinations (S.C.C.) in $X = T[G, (\mathcal{S}_{n_j}, \frac{1}{m_j})_j, D]$ are defined through the basic special convex combinations introduced in chapter 3. The main result of this chapter is the existence of seminormalized S.C.C. (Proposition 5.4) in every block subspace of X and Proposition 5.3 which asserts that the value of a functional with small weight acting on a large S.C.C. is small. Both will be used in the proof of the H.I. property of X. We also prove Proposition 5.5, which will be used in the proof of the H.I. property of the predual X_* of X.

Throughout this chapter we denote by X the space $T[G, (\mathcal{S}_{n_j}, \frac{1}{m_j})_j, D]$ where G is a ground set $(m_j)_{j\in\mathbb{N}}$, $(n_j)_{j\in\mathbb{N}}$ are the sequences defined in Notation 3.12 and D is a tree complete subset of the set $W[G, (\mathcal{S}_{n_j}, \frac{1}{m_j})_j]$.

DEFINITION 5.1 (special convex combinations). Let $\varepsilon > 0$, $j \in \mathbb{N}$ with $j > 1$ and let $(x_k)_{k\in\mathbb{N}}$ be a block sequence of the standard basis. A convex combination $\sum_{k\in F} a_k x_k$ of the sequence $(x_k)_{k\in\mathbb{N}}$ is said to be an (ε, j) *special convex combination* ((ε, j) S.C.C.) of $(x_k)_{k\in\mathbb{N}}$ if $\sum_{k\in F} a_k e_{t_k}$ (where $t_k = \min\operatorname{supp} x_k$ for each k) is an (ε, j) basic special convex combination i.e.

(1) $\{t_k : k \in F\} \in \mathcal{S}_{n_j}$
(2) For every $P \in \mathcal{S}_{2\log_2(m_j)(n_{j-1}+1)}$ we have that $\sum_{k:\, t_k\in P} a_k < \varepsilon$.

If moreover the sequence $(x_k)_{k\in\mathbb{N}}$ is such that $\|x_k\| \leq 1$ for all k and $\|\sum_{k\in F} a_k x_k\| \geq \frac{1}{2}$ we say that $\sum_{k\in F} a_k x_k$ is a *seminormalized* (ε, j) *special convex combination* of $(x_k)_{k\in\mathbb{N}}$.

REMARK 5.2. Let $(x_k)_{k\in\mathbb{N}}$ be a block sequence in X, $\varepsilon > 0$, $j \in \mathbb{N}$ with $j > 1$. Then for every $q \in \mathbb{N}$ such that $2\log_2(m_j)(n_{j-1} + 1) < q \leq n_j$ there exists an (ε, j) S.C.C. $\sum_{k\in F} a_k x_k$ with the additional property $\{\min\operatorname{supp} x_k : k \in F\} \in \mathcal{S}_q$. (If $t_k = \min\operatorname{supp} x_k$ for $k = 1, 2, \ldots$ then, from Remark 3.18, there exists an (ε, j) B.S.C.C. $\sum_{k\in F} a_k e_{t_k}$ with $\{t_k : k \in F\} \in \mathcal{S}_q$. Thus $\sum_{k\in F} a_k x_k$ is the desired (ε, j) S.C.C.)

PROPOSITION 5.3. Let $(x_k)_{k\in\mathbb{N}}$ be a block sequence in X with $\|x_k\| \leq C$ for each k, $\varepsilon > 0$ and $j \in \mathbb{N}$ with $j > 1$ such that $\varepsilon \leq \frac{1}{m_j}$, and let $\sum_{k\in F} a_k x_k$ be an (ε, j)

S.C.C. Then for every $i < j$ and $f \in A \cap D$ with $w(f) = m_i$ we have that

$$|f(\sum_{k\in F} a_k x_k)| \leq \frac{2C}{m_i}.$$

Proof. Let $f_1 < f_2 < \cdots < f_d$ be an $\mathcal{S}_{n_i}$ admissible sequence in D such that

$$f = \frac{1}{m_i}(f_1 + f_2 + \cdots + f_d) \in D.$$

It is enough to show that

$$|(\sum_{l=1}^{d} f_l)(\sum_{k\in F} a_k x_k)| \leq 2C. \tag{12}$$

We set

$$F_1 = \Big\{k \in F : \#\{l : \operatorname{supp} f_l \cap \operatorname{ran} x_k \neq \emptyset\} \leq 1\Big\} \text{ and } F_2 = F \setminus F_1.$$

For $k \in F_1$ we denote by l_k the unique $l \in \{1, 2, \ldots, d\}$ such that $\operatorname{supp} f_l \cap \operatorname{ran} x_k \neq \emptyset$, if such an l exists, and $f_{l_k} = 0$ in the opposite case. We have

$$|(\sum_{l=1}^{d} f_l)(\sum_{k\in F_1} a_k x_k)| = |\sum_{k\in F_1} a_k f_{l_k}(x_k)| \leq C \sum_{k\in F_1} a_k \leq C. \tag{13}$$

We set $t_k = \min\operatorname{supp} x_k$ for each k and observe that $\min\operatorname{supp} x_1 \geq 2$. For each $k \in F_2$ we choose $r_k \in \{1, 2, \ldots, d\}$ such that

$$\min\operatorname{supp} x_k \leq \min\operatorname{supp} f_{r_k} \leq \max\operatorname{supp} x_k.$$

Since $\{\min\operatorname{supp} f_l : l = 1, 2, \ldots, d\} \in \mathcal{S}_{n_i}$ it follows that $\{\min\operatorname{supp} f_{r_k} : k \in F_2\} \in \mathcal{S}_{n_i}$ thus $\{t_k : k \in F_2 \setminus \{\min F_2\}\} \in \mathcal{S}_{n_i}$. So the set $\{t_k : k \in F_2\}$ can be written as the union of a singleton and an element of $\mathcal{S}_{n_i}$. Since $n_i + 1 \leq 2\log_2(m_j)(n_{j-1}+1)$ we get that $\{t_k : k \in F_2\} \in \mathcal{S}_{2\log_2(m_j)(n_{j-1}+1)}$. The fact that $\sum_{k\in F} a_k x_k$ is an (ε, j) S.C.C. yields that $\sum_{k\in F_2} a_k < \varepsilon$.

Since $|f(x_k)| \leq \|x_k\| \leq C$ for each k we get that

$$|(\sum_{l=1}^{d} f_l)(\sum_{k\in F_2} a_k x_k)| \leq m_i \sum_{k\in F_2} a_k |f(x_k)| \leq m_i \varepsilon C \leq m_i \frac{1}{m_j} C < C. \tag{14}$$

From (13) and (14) we conclude that

$$|(\sum_{l=1}^{d} f_l)(\sum_{k\in F} a_k x_k)| \leq |(\sum_{l=1}^{d} f_l)(\sum_{k\in F_1} a_k x_k)| + |(\sum_{l=1}^{d} f_l)(\sum_{k\in F_2} a_k x_k)| < C + C = 2C. \tag{15}$$

and inequality (12) has been proved. □

The next proposition, which ensures the existence of seminormalized S.C.C. in every block subspace, is contained in [**6**]. For the sake of completeness we include the proof here.

PROPOSITION 5.4. *Let $(x_k)_{k\in\mathbb{N}}$ be a normalized block sequence in X, $\varepsilon > 0$ and $j \in \mathbb{N}$ with $j > 1$ such that D is j closed (Definition 3.8(iv)). Then there exists a block sequence $(y_k)_{k\in\mathbb{N}}$ of $(x_k)_{k\in\mathbb{N}}$ with $\|y_k\| \leq 1$ for all k and an (ε, j) S.C.C. $\sum_{k\in F} a_k y_k$ such that*

(1) $\|\sum_{k\in F} a_k y_k\| \geq \frac{1}{2}$.
(2) The family $\{y_k : k \in F\}$ is $\mathcal{S}_{2\log_2(m_j)(n_{j-1}+1)+1}$ admissible.

Proof. Assume the contrary. Set $r = 2\log_2(m_j)(n_{j-1}+1)+1$ and $m = \log_2(m_j)$ and observe that $m > 1$ and $mr < n_j$ (see Notation 3.12). We will make an inductive construction and arrive to a contradiction after m inductive steps. We thus construct block sequences $(y_k^l)_{k\in\mathbb{N}}$ for $0 \leq l \leq m$ satisfying the following conditions:

(i) For $1 \leq l \leq m$, $(y_k^l)_{k\in\mathbb{N}}$ is a block sequence of $(y_k^{l-1})_{k\in\mathbb{N}}$ and each y_k^l is a convex combination of the sequence $(x_k)_{k\in\mathbb{N}}$.
(ii) For all l and k the family $\{\operatorname{supp} x_i : \operatorname{supp} x_i \subset \operatorname{supp} y_k^l\}$ is $\mathcal{S}_{lr}$ admissible.
(iii) $\frac{1}{m_j} \leq \|y_k^l\| < \frac{1}{2^l}$ for $1 \leq l \leq m$ and all k.

When the inductive construction is complete (iii) yields that $\frac{1}{m_j} < \frac{1}{2^m}$, a contradiction since $m = \log_2(m_j)$.

We proceed to the inductive construction. We set $y_k^0 = x_k$ for all k. To treat the general inductive step let $1 \leq l \leq m$ and suppose that (y_k^{l-1}) has already been defined satisfying the inductive assumptions. Using Remark 5.2 we choose a block sequence $(y_k^l)_{k\in\mathbb{N}}$ of $(y_k^{l-1})_{k\in\mathbb{N}}$ such each y_k^l is an (ε, j) S.C.C. of $(y_k^{l-1})_{k\in\mathbb{N}}$ with the family $\{\operatorname{supp} y_i^{l-1} : \operatorname{supp} y_i^{l-1} \subset y_k^l\}$ being $\mathcal{S}_r$ admissible. Since by our inductive assumptions for each k the family $\{\operatorname{supp} x_i : \operatorname{supp} x_i \subset \operatorname{supp} y_k^{l-1}\}$ is $\mathcal{S}_{(l-1)r}$ admissible it follows that the family $\{\operatorname{supp} x_i : \operatorname{supp} x_i \subset \operatorname{supp} y_k^l\}$ is $\mathcal{S}_r * \mathcal{S}_{(l-1)r} = \mathcal{S}_{lr}$ admissible (and thus $\mathcal{S}_{n_j}$ admissible). It also follows from our inductive assumptions that each y_k^l is a convex combination of $(x_k)_{k\in\mathbb{N}}$ and hence from Remark 3.10 we get that $\|y_k^l\| \geq \frac{1}{m_j}$. If $\|y_k^l\| \geq \frac{1}{2^l}$ for some k and $y_k^l = \sum_{i\in F} a_i y_i^{l-1}$ is the expansion of y_k^l as a convex combination of the sequence $(y_k^{l-1})_{k\in\mathbb{N}}$ we get that $\|\sum_{i\in F} a_i(2^{l-1}y_i^{l-1})\| \geq \frac{1}{2}$ which, in conjunction to our inductive condition that $\|y_i^{l-1}\| < \frac{1}{2^{l-1}}$ for all i, contradicts our assumption that the proposition does not hold. Therefore $\|y_k^l\| \leq \frac{1}{2^l}$ for all k.

The proof of the proposition is complete. □

The statement of the next proposition, as well as its proof, is the analogue of Proposition 5.4 in the predual space X_*.

PROPOSITION 5.5. Let $(f_k)_{k\in\mathbb{N}}$ be a block sequence of the basis $(e_k^*)_{k\in\mathbb{N}}$ of X_* such that $f_k \in D$ and $\|f_k\| \geq \frac{1}{2}$ for all k. Let also $j \in \mathbb{N}$, $j > 1$ such that D is j closed (Definition 3.8(iv)). Then there exists a finite block sequence $(h_k)_{k\in F}$ of $(f_k)_{k\in\mathbb{N}}$ such that

(1) $\|h_k\| > \frac{1}{2}$ for all k and $\|\sum_{k\in F} h_k\| \leq 1$.
(2) The set $\{\min\operatorname{supp} h_k : k \in F\}$ is a maximal $\mathcal{S}_{2\log_2(m_j)(n_{j-1}+1)+1}$ set.
(3) There exists a $c < 1$ such that each h_k is of the form $h_k = c\sum_{i\in F_k} f_i$.

Proof. Assume the contrary. We set $r = 2\log_2(m_j)(n_{j-1}+1)+1$ and $m = \log_2(m_j)+1$ and observe that $m > 2$ and $mr < n_j$ (Notation 3.12). Our intention is to make an inductive construction in X_*, similar to that made in the proof of Proposition 5.4 in X, and arrive to a contradiction after m inductive steps. We proceed by constructing block sequences $(h_k^l)_{k\in\mathbb{N}}$ for $0 \le l \le m$ satisfying the following conditions:

(i) For all l and k, h_k^l is of the form $h_k^l = \sum_{i\in F_k^l} f_i$ and the set $\{\min\operatorname{supp} f_i : i \in F_k^l\}$ is a maximal $\mathcal{S}_{lr}$ set.

(ii) $2^{l-1} < \|h_k^l\| \le m_j$ for $1 \le l \le m$ and all k.

After the completion of the inductive construction, (ii) yields that $2^{m-1} < m_j$ which contradicts to the choice of m.

We set $h_k^0 = f_k$ for all k. Suppose that the sequence $(h_k^{l-1})_{k\in\mathbb{N}}$ has been defined satisfying the inductive assumptions. We consider a partition $(H_k^l)_{k\in\mathbb{N}}$ of the natural numbers into successive finite sets such that for each k the set $\{\min\operatorname{supp} h_i^{l-1} : i \in H_k^l\}$ is an $\mathcal{S}_r$ maximal set. We set $h_k^l = \sum_{i\in H_k^l} h_i^{l-1}$ for all k. By our inductive assumption (i) and the definition of h_k^l we get that each h_k^l may written as $h_k^l = \sum_{i\in F_k^l} f_i$ with the set $\{\min\operatorname{supp} f_i : i \in F_k^l\}$ being $\mathcal{S}_{lr}$ maximal. Since $lr < n_j$ this also gives that $\{f_i : i \in F_k^l\}$ is an $\mathcal{S}_{n_j}$ admissible family in D. Thus our assumption that the norming set D is j closed yields that $\|h_k^l\| \le m_j$. On the other hand, if $\|h_k^l\| \le 2^{l-1}$ for some k, we get that $\|\sum_{i\in H_k^l}(\frac{1}{2^{l-1}}h_i^{l-1})\| \le 1$. By our inductive condition (ii) we also have that $\|\frac{1}{2^{l-1}}h_i^{l-1}\| > \frac{1}{2}$ for all i. This leads to a contradiction since we have assumed that the proposition does not hold. Therefore $\|h_k^l\| > 2^{l-1}$ for each k.

The proof of the proposition is complete. □

CHAPTER 6

Rapidly increasing sequences

Rapidly increasing sequences (R.I.S.) are another crucial component for showing the H.I. property of the examples presented here. They were defined for similar reasons in [**47**] and in the frame of H.I. spaces in [**28**]. Roughly speaking, rapidly increasing sequences are seminormalized block sequences such that each functional norming one of the terms of the sequence, becomes negligible in the rest. This forces certain special convex combinations of the R.I.S. to behave similarly to the corresponding basic special convex combinations.

DEFINITION 6.1. Let $C \geq 1$ and $\varepsilon > 0$. A block sequence $(x_k)_{k\in\mathbb{N}}$ in X is said to be a (C,ε) *rapidly increasing sequence* ((C,ε) R.I.S.) if the following conditions hold

(1) The sequence $(x_k)_{k\in\mathbb{N}}$ is (G,ε) Schreier with $\|x_k\| \leq C$ for all k and $\min\operatorname{supp} x_1 \geq 7$.
(2) There exists a strictly increasing sequence $(j_k)_{k\in\mathbb{N}}$ of integers such that:
   (a) $(\max\operatorname{supp} x_k)\frac{1}{\varepsilon} \leq m_{j_{k+1}}$ for each k.
   (b) For each $k = 1,2,\ldots$ and $h \in A \cap D$ with $w(h) < m_{j_k}$ we have $|h(x_k)| \leq \dfrac{C}{w(f)}$.

In other words the sequence $(x_k)_{k\in\mathbb{N}}$ is said to be a (C,ε) R.I.S. if it satisfies the assumptions of the basic inequality (Proposition 4.3) with constants C and ε.

PROPOSITION 6.2. Let $(x_k)_{k\in\mathbb{N}}$ be a (C,ε) rapidly increasing sequence in X (where $C \geq 1$, $\varepsilon > 0$) and $\delta > 0$, $j \in \mathbb{N}$ with $j > 1$ such that $\max\{\delta,\varepsilon\} \leq \dfrac{1}{m_j}$. Then:

(1a) If $\sum_{k\in F} a_k x_k$ is a (δ,j) S.C.C. of the sequence $(x_k)_{k\in\mathbb{N}}$ then

$$\|\sum_{k\in F} a_k x_k\| \leq \frac{5C}{m_j}.$$

(1b) If moreover $\max\{\delta,\varepsilon\} \leq \dfrac{1}{m_j^2}$ then for every $f \in A \cap D$ with $w(f) < m_j$ we have that

$$|f(\sum_{k\in F} a_k x_k)| \leq \frac{7C}{w(f)m_j}.$$

(2) If $\max\{\delta,\varepsilon\} \leq \dfrac{1}{m_j^2}$, the sequence $(a_k)_{k\in F}$ is such that $\sum_{k\in F} |a_k| x_k$ is a (δ,j) S.C.C. and also if the assumption (d) of the basic inequality (Proposition 4.3) is fulfilled for $j_0 = j$ and for the linear combination $\sum_{k\in F} a_k x_k$ of

$(x_k)_{k\in\mathbb{N}}$ then

$$\|\sum_{k\in F} a_k x_k\| \leq \frac{5C}{m_j^2}.$$

Proof. For the proof of (1a) and (1b) we consider an $f \in A \cap D$. It follows from the basic inequality (Proposition 4.3) that there exists an $h_1 \in \mathrm{conv}_{\mathbb{Q}}\{h \in B : w(h) = w(f)\}$, a $t \in \mathbb{N}$ with $t \notin \operatorname{supp} h_1$ and an $h_2 \in c_{00}$ with $\|h_2\|_\infty \leq \varepsilon$ such that

$$(16) \qquad |f(\sum_{k\in F} a_k x_k)| \leq C(2(h_1 + e_t^*) + h_2)(\sum_{k\in F} a_k e_{t_k})$$

where $t_k = \min\operatorname{supp} x_k$ for each k. From Lemma 3.19(i) we get that

$$\begin{aligned} |f(\sum_{k\in F} a_k x_k)| &\leq C(2h_1(\sum_{k\in F} a_k e_{t_k}) + 2\max_{k\in F} a_k + \varepsilon) \\ &\leq C(2\frac{1}{m_j} + 2\delta + \varepsilon) \leq \frac{5C}{m_j}. \end{aligned}$$

We conclude that

$$\|\sum_{k\in F} a_k x_k\| \leq \frac{5C}{m_j}.$$

Suppose now that $\max\{\delta, \varepsilon\} \leq \frac{1}{m_j^2}$ and $w(f) < m_j$. It follows from relation (16) and Lemma 3.19(i) that

$$\begin{aligned} |f(\sum_{k\in F} a_k x_k)| &\leq C(2h_1(\sum_{k\in F} a_k e_{t_k}) + 2\max_{k\in F} a_k + \varepsilon) \\ &\leq C(2\frac{2}{w(f)m_j} + 2\delta + \varepsilon) \leq C(\frac{4}{w(f)m_j} + 2\frac{1}{m_j^2} + \frac{1}{m_j^2}) \\ &\leq \frac{7C}{w(f)m_j}. \end{aligned}$$

which completes the proof of (1b).

We pass to the proof of (2). Let $f \in A \cap D$. Assume first that $w(f) \neq m_j$. Since the linear combination $\sum_{k\in F} a_k x_k$ satisfies assumption (d) of the basic inequality (Proposition 4.3) for $j_0 = j$ we may choose h_1, h_2, t satisfying inequality (16), with h_1 satisfying the additional property that can be written in the form $h_1 = \sum r_i \tilde{h}_i$ with $r_i \in \mathbb{Q}^+$, $\sum r_i = 1$ and for each i the functional $\tilde{h}_i$ belongs to B and has a tree $T_{\tilde{h}_i} = (f_a^i)_{a\in\mathcal{C}_i}$ with $w(f_a^i) \neq m_j$ for all $a \in \mathcal{C}_i$.

From Lemma 3.19(ii) we conclude that

$$\begin{aligned} |f(\sum_{k\in F} a_k x_k)| &\leq C(2h_1(\sum_{k\in F} |a_k| e_{t_k}) + 2\max_{k\in F} |a_k| + \varepsilon) \\ &\leq C(2\frac{1}{m_j^2} + 2\delta + \varepsilon) \leq \frac{5C}{m_j^2}. \end{aligned}$$

The case $w(f) = m_j$ follows immediately from the assumption of condition (d) of the basic inequality (Proposition 4.3) and the fact that $\max_{k\in F} |a_k| \leq \delta \leq \frac{1}{m_j^2}$ and $\varepsilon \leq \frac{1}{m_j^2}$.

This completes the proof of (2). $\square$

We proceed with the next definition.

DEFINITION 6.3. A ground set G (Definition 3.1) is said to be $\mathcal{S}_2$ *bounded* provided that the ground space Y_G, satisfies the following conditions:

(i) The space ℓ^1 does not embed into Y_G.

(ii) For every weakly null block sequence $(y_n)_{n\in\mathbb{N}}$ in Y_G and every $\varepsilon > 0$ there exists a subsequence $(y_n)_{n\in M}$ such that for every $g \in G$ the set

$$\{\min\operatorname{supp} y_n : n \in M \text{ and } |g(y_n)| > \varepsilon\}$$

belongs to $\mathcal{S}_2$, the second Schreier family (i.e. the sequence $(y_n)_{n\in M}$ is (G,ε) Schreier).

REMARKS 6.4. 1. Proposition 3.11(i) yields that every bounded block sequence $(x_k)_{k\in\mathbb{N}}$ in $X = T[G, (\mathcal{S}_{n_j}, \frac{1}{m_j})_j, D]$ is also a bounded block sequence in Y_G. Assuming additionally that the set G is $\mathcal{S}_2$ bounded we have that the space Y_G contains no isomorphic copy of ℓ^1. Hence, using Rosenthal's ℓ^1 theorem [**46**], we conclude that every bounded block sequence $(x_k)_{k\in\mathbb{N}}$ in X has a subsequence $(x_{l_k})_{k\in\mathbb{N}}$ which is weakly Cauchy in Y_G. Therefore the sequence of successive differences of this sequence (i.e. the sequence $(y_k)_{k\in\mathbb{N}}$ defined by $y_k = x_{l_{2k-1}} - x_{l_{2k}}$) is a weakly null block sequence in Y_G. It follows that for every $\varepsilon > 0$ there exists a (G,ε) Schreier subsequence $(y_k)_{k\in M}$.
2. The ground set $G_0 = \{\pm e_q : q \in \mathbb{N}\}$ is clearly $\mathcal{S}_2$ bounded.
3. If G is an $\mathcal{S}_2$ bounded ground set, $(y_n)_{n\in\mathbb{N}}$ is a weakly null block sequence in Y_G and $\varepsilon > 0$ then there exists a subsequence $(y_n)_{n\in M}$ such that $\{\min\operatorname{supp} y_n : n \in M \text{ and } |g(y_n)| > \varepsilon\} \in \mathcal{S}_2$ for every $g \in \overline{G}^{w^*}$.

LEMMA 6.5. Let G be an $\mathcal{S}_2$ bounded set, $(x_k)_{k\in\mathbb{N}}$ a block sequence in $X = T[G, (\mathcal{S}_{n_j}, \frac{1}{m_j})_j, D]$ and $\varepsilon > 0$. Suppose also that there exists a strictly increasing sequence $(i_k)_{k\in\mathbb{N}}$ of the natural numbers such that each x_k is a seminormalized $(\frac{1}{m_{i_k}}, i_k)$ S.C.C. Then there exists a subsequence $(x_{l_k})_{k\in\mathbb{N}}$ of $(x_k)_{k\in\mathbb{N}}$ such that the sequence $(y_k)_{k\in\mathbb{N}}$ of successive differences of this sequence (i.e. $y_k = x_{l_{2k-1}} - x_{l_{2k}}$) is a $(4,\varepsilon)$ rapidly increasing sequence.

Proof. We may clearly assume, passing to a subsequence, that $\min\operatorname{supp} x_1 \geq 7$ and that $(\max\operatorname{supp} x_k)\frac{1}{\varepsilon} \leq m_{i_{k+1}}$ for all k. Since the ground set G is $\mathcal{S}_2$ bounded, there exists a subsequence $(x_{l_k})_{k\in\mathbb{N}}$ of $(x_k)_{k\in\mathbb{N}}$ such that the sequence $(y_k)_{k\in\mathbb{N}}$ with $y_k = x_{l_{2k-1}} - x_{l_{2k}}$ is (G,ε) Schreier. It follows from the bimonotonicity of $(e_n)_{n\in\mathbb{N}}$ and the triangle inequality that $\frac{1}{2} \leq \|y_k\| \leq 2$ for all k. We set

$$j_k = i_{l_{2k-1}} \text{ for } k = 1, 2, \ldots$$

For each k we have that

$$m_{j_{k+1}} = m_{i_{l_{2k+1}}} \geq m_{i_{l_{2k}+1}} \geq (\max\operatorname{supp} x_{l_{2k}})\frac{1}{\varepsilon} = (\max\operatorname{supp} y_k)\frac{1}{\varepsilon}.$$

For every $f \in A \cap D$ with $w(f) < m_{j_k} (= m_{i_{l_{2k-1}}} < m_{i_{l_{2k}}})$ we also have that

$$|f(y_k)| \leq |f(x_{l_{2k-1}})| + |f(x_{l_{2k}})| < \frac{2}{w(f)} + \frac{2}{w(f)} = \frac{4}{w(f)}.$$

The last inequality follows from Proposition 5.3 since for each k the vector x_k is a $(\frac{1}{m_{i_k}}, i_k)$ s.c.c of a sequence $(z_i)_{i\in\mathbb{N}}$ with $\|z_i\| \leq 1$. □

DEFINITION 6.6. Let $\lambda > 1$. A Banach space $(X, \|\ \|)$ is said to be λ-distortable if there exists an equivalent norm $|\ |$ on X such that for every infinite dimensional subspace Y of X

$$\sup\Big\{\frac{|x|}{|y|} : x, y \in Y \text{ and } \|x\| = \|y\| = 1\Big\} \geq \lambda.$$

The space $(X, \|\ \|)$ is said to be arbitrarily distortable if it is λ-distortable for all $\lambda > 1$.

PROPOSITION 6.7. If the ground set G is $\mathcal{S}_2$ bounded then for every tree complete set $D \subset W[G, (\mathcal{S}_{n_j}, \frac{1}{m_j})_j]$ the space $X = T[G, (\mathcal{S}_{n_j}, \frac{1}{m_j})_j, D]$ is arbitrarily distortable.

Proof. The proof is obtained following the same lines with the proof of Corollary 2.15 of [**6**]. □

Proposition 6.7 yields the following.

COROLLARY 6.8. For every $\mathcal{S}_2$ bounded ground set G and tree complete set $D \subset W[G, (\mathcal{S}_{n_j}, \frac{1}{m_j})_j]$ the space $X = T[G, (\mathcal{S}_{n_j}, \frac{1}{m_j})_j, D]$ does not contain isomorphically ℓ^1.

REMARK 6.9. In the space Y_G (reps. $X = T[G, (\mathcal{S}_{n_j}, \frac{1}{m_j})_j, D]$) the w^* closure of G considered as a subset of $B_{Y_G^*}$ (resp. B_{X^*}) coincides with the closure of G in the pointwise topology of $\ell^\infty(\mathbb{N})$.

The next proposition concerns the relation of the ground set G to the dual space X^*.

PROPOSITION 6.10. If G is $\mathcal{S}_2$ bounded and $D \subset W[G, (\mathcal{S}_{n_j}, \frac{1}{m_j})_j]$ is tree complete then the dual X^* of the space $X = T[G, (\mathcal{S}_{n_j}, \frac{1}{m_j})_j, D]$ is the norm closed linear span of the w^* closure of G.

$$X^* = \overline{\text{span}}(\overline{G}^{w^*}).$$

Proof. Let $Z = \overline{\text{span}}(\overline{G}^{w^*})$ Obviously $Z \subset X^*$. Suppose that $Z \neq X^*$ and select $x^* \in X^* \setminus Z$ with $\|x^*\| = 1$ and $x^{**} \in X^{**}$ such that $\|x^{**}\| = 1$, $x^{**}(x^*) = 1$ and $Z \subset \operatorname{Ker} x^{**}$.

Since X does not contain isomorphically ℓ^1, Odell-Rosenthal's Theorem [**42**] yields a sequence $(x_n)_{n\in\mathbb{N}}$ in X, w^* converging to x^{**}. Since $e_k^* \in G \subset Z$ for all k and $Z \subset \operatorname{Ker} x^{**}$ by a sliding hump argument we may assume that $(x_n)_{n\in\mathbb{N}}$ is a block sequence; we may also assume that $\|x_n\| = 1$ for all n. Since $x^*(x_n) \longrightarrow x^{**}(x^*) = 1$ we may also assume that $x^*(x_n) > \frac{1}{2}$ for all n. Select $j \in \mathbb{N}$ with $m_j > 20$ and set $\varepsilon = \frac{1}{m_j}$. We select a strictly increasing sequence $(i_k)_{k\in\mathbb{N}}$ of natural numbers and a block sequence $(y_k)_{k\in\mathbb{N}}$ in X such that

(1) Each y_k is a $(\frac{1}{m_{i_k}}, i_k)$ S.C.C. of the sequence $(x_n)_{n\in\mathbb{N}}$.

(2) $\min\operatorname{supp} y_1 \geq 7$.

(3) $m_{i_{k+1}} > \max\operatorname{supp} y_k \cdot \frac{1}{\varepsilon} \quad k = 1, 2, \ldots$

Clearly $\frac{1}{2} \leq \|y_k\| \leq 1$ for all k. Moreover, since for every $g \in \overline{G}^{w^*}$ we have that $x^{**}(g) = 0$ hence $g(x_n) \overset{n\to\infty}{\longrightarrow} 0$ and $(y_k)_{k\in\mathbb{N}}$ is a convex block sequence of $(x_n)_{n\in\mathbb{N}}$ we get that $\lim_k g(y_k) = 0$ for every such a g.

Now Rainwater's Theorem [**45**] yields that $(y_k)_{k\in\mathbb{N}}$ is a weakly null sequence in the space Y_G. Since G is $\mathcal{S}_2$ bounded, we may assume, passing to a subsequence, that the sequence $(y_k)_{k\in\mathbb{N}}$ is (G, ε) Schreier. Proposition 5.3 yields that for $f \in D$ with $w(f) < m_{i_k}$ we have $|f(y_k)| \leq \frac{2}{w(f)}$. We have proved up to this point that $(y_k)_{k\in\mathbb{N}}$ is a $(2, \varepsilon)$ R.I.S.

Now let $x = \sum_{k\in F} a_k y_k$ be an (ε, j) S.C.C. of the sequence $(y_k)_{k\in\mathbb{N}}$. Since x is also a convex combination of the sequence $(x_n)_{n\in\mathbb{N}}$ and $x^*(x_n) > \frac{1}{2}$ for all n it follows that $\|x\| \geq \frac{1}{2}$. On the other hand Proposition 6.2(1a) yields that $\|x\| \leq \frac{5\cdot 2}{m_j} < \frac{1}{2}$, a contradiction.

Therefore $X^* = \overline{\operatorname{span}}(\overline{G}^{w^*})$. □

CHAPTER 7

Defining D to obtain a H.I. space X_G

This chapter contains one of the main results of this paper. Namely we show that for a ground set G which is $\mathcal{S}_2$ bounded and for an appropriate norming set D the space $X = T[G,(\mathcal{S}_{n_j},\frac{1}{m_j})_j, D]$ is H.I. The first part is devoted to the definition of the set D. As we have noted before, we choose D to be closed under rational convex combinations in order to force the predual X_* of X to be H.I. Then, using the results of the previous chapters, we prove that X is indeed a H.I. space.

DEFINITION 7.1 (The norming set D). Let $\mathcal{W} = \Big\{(f_1, f_2, \ldots, f_l) : f_i \in c_{00}, f_i \neq 0, f_i(n) \in \mathbb{Q}\ \forall n \in \mathbb{N},\ \forall i = 1,2,\ldots l,\ f_1 < f_2 < \cdots < f_l,\ l \in \mathbb{N}\Big\}$. Since $\mathcal{W}$ is countable we can define a Gowers-Maurey type injection $\sigma : \mathcal{W} \longrightarrow \{2j : j \in \mathbb{N}\}$ such that

$$m_{\sigma(f_1,\ldots,f_l)} > \max\Big\{\frac{1}{|f_i(e_k)|} : k \in \operatorname{supp} f_i,\ i = 1,\ldots,l\Big\} \cdot \max\operatorname{supp} f_l$$

We consider the set $W = W\left[G,\left(\mathcal{S}_{n_j},\frac{1}{m_j}\right)_j\right]$ and the subset $A \subset W$ as defined in Definition 3.3. A finite (resp. infinite) sequence $(f_i)_{i=1,\ldots,k}$ (resp. $(f_i)_{i\in\mathbb{N}}$) of successive functionals in A is said to be a *special sequence* if

(i) $w(f_1) \in \{m_{2j} : j \in \mathbb{N}\}$
(ii) For each $i = 1,\ldots,k-1$ (resp. $i \in \mathbb{N}$) $w(f_{i+1}) = m_{\sigma(f_1,\ldots,f_i)}$.

We recursively construct sequences $(C_n)_{n\in\mathbb{N}}$, $(D_n)_{n\in\mathbb{N}}$ of subsets of c_{00} as follows:

$$C_0 = G, \qquad D_0 = \operatorname{conv}_{\mathbb{Q}}(C_0)$$

$$\begin{aligned} C_{n+1} = {} & C_n \cup \bigcup_{j=1}^{\infty}\Big\{\frac{1}{m_{2j}}\sum_{i=1}^{d} f_i : f_i \in D_n \text{ for each } i = 1,2,\ldots,d \\ & \text{and the sequence } (f_1,\ldots,f_d) \text{ is } \mathcal{S}_{n_{2j}} \text{ admissible}\Big\} \\ & \cup \bigcup_{j=1}^{\infty}\Big\{\pm E\Big(\frac{1}{m_{2j-1}}\sum_{i=1}^{d} f_i\Big) : (f_1,\ldots,f_d) \text{ is an } \mathcal{S}_{n_{2j-1}} \text{ admissible special} \\ & \text{sequence in } C_n \text{ with } w(f_1) > m_{2j-1} \text{ and } E \text{ is an interval of } \mathbb{N}\Big\} \end{aligned}$$

$$D_{n+1} = \operatorname{conv}_{\mathbb{Q}}(C_{n+1})$$

We set

$$C = \bigcup_{n=0}^{\infty} C_n \text{ and } D = \bigcup_{n=0}^{\infty} D_n.$$

REMARKS 7.2. (i) It is easy to see that the set D is tree complete. Indeed, clearly $G \subset D$, for every $f \in D$ there exists a tree $T_f = (f_a)_{a\in\mathcal{A}}$ of f with $f_a \in D$ for all $a \in \mathcal{A}$ and D is $2j$ closed for each $j \in \mathbb{N}$. As also can be easily seen by induction on n, for every $f \in D_n$ and every interval E of the natural numbers we have that $Ef \in D$ and $-f \in D$.
(ii) The set D is rationally convex, i.e. $\text{conv}_{\mathbb{Q}}(D) = D$, thus since $B_{X^*} = \overline{\text{conv}(\mathcal{D})}^p$ it follows that $B_{X^*} = \overline{D}^p$.

In the sequel, for a ground set G which is $\mathcal{S}_2$ bounded, we denote by X_G the space $X_G = T\left[G, \left(\mathcal{S}_{n_j}, \frac{1}{m_j}\right)_j, D\right]$ where D is the set determined by G as it is above defined.

THEOREM 7.3. *If G is $\mathcal{S}_2$ bounded then the space $X_G = T[G, (\mathcal{S}_{n_j}, \frac{1}{m_j})_j, D]$ is Hereditarily Indecomposable.*

To prove this theorem we need some preparatory work. We give a definition and prove a proposition.

DEFINITION 7.4. A double sequence $(x_k, x_k^*)_{k\in\mathbb{N}}$ is said to be a $(\theta, 2j+1)$ *dependent couple* (for $j \in \mathbb{N}$ and $0 \le \theta \le 1$) if there exists a sequence $(2j_k)_{k\in\mathbb{N}}$ of even integers such that the following conditions are fulfilled:
(i) $(x_k)_{k\in\mathbb{N}}$ is a block sequence in X_G with $\min\operatorname{supp} x_1 > 3m_{2j+1}^2$, and for each k, $x_k = \theta_k m_{2j_k} \sum_{i\in F_k} a_i^k x_i^k$ where $\sum_{i\in F_k} a_i^k x_i^k = a(n_{2j_k}, (x_i^k)_i)$ is a $(\frac{1}{4m_{2j_k}^2}, 2j_k)$ S.C.C. of $(x_i^k)_i$, the sequence $(x_i^k)_i$ is a $(4, \frac{1}{m_{2j_k}^2})$ R.I.S. and $\theta_k \le 1$.
(ii) $(x_k^*)_{k\in\mathbb{N}}$ is a special sequence with $w(x_k^*) = m_{2j_k}$ and $2j_1 \ge 2j+4$.
(iii) $x_k^*(x_k) = \theta$ and $\operatorname{ran} x_k^* = \operatorname{ran} x_k$ for all k.

PROPOSITION 7.5. Let $(x_k, x_k^*)_{k\in\mathbb{N}}$ be a $(\theta, 2j+1)$ dependent couple. Let also $a(n_{2j+1}, (x_k)_{k\in\mathbb{N}}) = \sum_{k=1}^n a_k x_k$. Then

$$\|\sum_{k=1}^n (-1)^k a_k x_k\| \le \frac{140}{m_{2j+1}^2}.$$

.

Proof. Let $(2j_k)_{k\in\mathbb{N}}$ be as in Definition 7.4. Our goal is to prove that the sequence $(x_k)_{k\in\mathbb{N}}$ is an $(28, \varepsilon)$ R.I.S. for $\varepsilon = \frac{1}{m_{2j+1}^2}$ and the linear combination $\sum_{k=1}^n (-1)^k a_k x_k$ also satisfies the assumption (d) of the basic inequality (Proposition 4.3) for $j_0 = 2j+1$. Then, $\sum_{k=1}^n a_k x_k$ is an $(\varepsilon, 2j+1)$ S.C.C. (since $\min\operatorname{supp} x_1 \ge 3m_{2j+1}^2$) thus Proposition 6.2(2) yields that

$$\|\sum_{k=1}^n (-1)^k a_k x_k\| \le \frac{5 \cdot 28}{m_{2j+1}^2} = \frac{140}{m_{2j+1}^2}.$$

Since the convex combination $\sum_{i\in F_k} a_i^k x_i^k = a(n_{2j_k},(x_i^k)_i)$ is a $(\frac{1}{4m_{2j_k}^2}, 2j_k)$ S.C.C. and $(x_i^k)_i$ is a $(4,\frac{1}{m_{2j_k}^2})$ R.I.S. it follows from Proposition 6.2(1a) that $\|\sum_{i\in F_k} a_i^k x_i^k\| \le \frac{5\cdot 4}{m_{2j_k}} = \frac{20}{m_{2j_k}}$. Hence

$$\|x_k\| = \theta_k m_{2j_k} \|\sum_{i\in F_k} a_i^k x_i^k\| \le 20$$

for all k.

In order to prove that the sequence $(x_k)_{k\in\mathbb{N}}$ is (G,ε) Schreier we will show that our assumptions imply something even stronger. Namely we have that $|g(x_k)| \le \varepsilon$ for every $g \in G$ and all k.

Indeed, let $g \in G$ and $k = 1,2,\ldots$. We set

$$I_k = \{i \in F_k : |g(x_i^k)| > \frac{1}{m_{2j_k}^2}\}.$$

The sequence $(x_i^k)_i$ is $(G,\frac{1}{m_{2j_k}^2})$ Schreier (from the definition of R.I.S.) thus the set $\{\min\operatorname{supp} x_i^k : i \in I_k\}$ belongs to $\mathcal{S}_2$ Since $\sum_{i\in F_k} a_i^k x_i^k$ is a $(\frac{1}{4m_{2j_k}^2}, 2j_k)$ S.C.C. it follows that $\sum_{i\in I_k} a_i^k < \frac{1}{4m_{2j_k}^2}$. So

$$|g(\sum_{i\in F_k} a_i^k x_i^k)| \le \sum_{i\in I_k} a_i^k |g(x_i^k)| + \sum_{i\in F_k\setminus I_k} a_i^k |g(x_i^k)| \le 4\sum_{i\in I_k} a_i^k + \frac{1}{m_{2j_k}^2} \le \frac{2}{m_{2j_k}^2}.$$

We finally get that

$$|g(x_k)| = \theta_k m_{2j_k} |g(\sum_{i\in F_k} a_i^k x_i^k)| \le \theta_k m_{2j_k} \frac{2}{m_{2j_k}^2} \le \frac{2}{m_{2j_k}} < \frac{1}{m_{2j+1}^2} = \varepsilon.$$

Our assumptions about the injection σ yield that for all k

$$m_{2j_{k+1}} = m_{\sigma(x_1^*,\ldots,x_k^*)} > \max\operatorname{supp} x_k^* \cdot m_{2j_k} > \max\operatorname{supp} x_k \cdot \frac{1}{\varepsilon}.$$

For every $k = 1,2,\ldots$ and $h \in A \cap D$ with $w(h) < m_{2j_k}$, Proposition 6.2(1b) yields that

$$|h(x_k)| = \theta_k m_{2j_k} |h(\sum_{i\in F_k} a_i^k x_i^k)| \le m_{2j_k} \frac{7\cdot 4}{w(h) m_{2j_k}} = \frac{28}{w(h)}.$$

We have shown up to this point that the sequence $(x_k)_{k\in\mathbb{N}}$ is a $(28,\varepsilon)$ R.I.S. and it remains to prove that assumption (d) of the basic inequality (Proposition 4.3) is fulfilled for $j_0 = 2j+1$ and the linear combination $\sum_{k=1}^{n} (-1)^k a_k x_k$.

Let $f \in A \cap D$ with $w(f) = m_{2j+1}$. Then f has the form

$$f = \frac{1}{m_{2j+1}} (F x_{t-1}^* + x_t^* + x_{t+1}^* + \cdots + x_{r-1}^* + f_r + f_{r+1} + \cdots + f_d)$$

where $x_1^*, x_2^*, \ldots, x_{r-1}^*, f_r, f_{r+1}, \ldots, f_d$ is an $\mathcal{S}_{n_{2j+1}}$ admissible special sequence with $x_r^* \neq f_r$, $w(x_r^*) = w(f_r)$ and F is an interval of the natural numbers of the form $[m, \max\operatorname{supp} x_{t-1}^*]$; we also recall that $w(x_1^*) = m_{2j_1} > m_{2j+1}$. (In the case $r = 1$ we have $x_1^* \neq f_1$ and $w(f_1) > m_{2j+1}$.)

We first estimate the quantity $f(x_k)$ for each k.

- For $k < t-1$ we have $f(x_k) = 0$.
- For $k = t-1$ we have $|f(x_k)| = \frac{1}{m_{2j+1}}|Fx^*_{t-1}(x_{t-1})| \leq \frac{1}{m_{2j+1}}\|x_{t-1}\| \leq \frac{20}{m_{2j+1}}$.
- For $k = t, \ldots, r-1$ we have that $|f(x_k)| = \frac{1}{m_{2j+1}}|x^*_k(x_k)| = \frac{\theta}{m_{2j+1}}$.
- For $k = r$, $|f(x_k)| \leq \frac{20}{m_{2j+1}} + \frac{1}{m_{2j+1}}|(f_{r+1} + \ldots + f_d)(x_r)|$.
- For $k > r$, $|f(x_k)| = \frac{1}{m_{2j+1}}|(f_r + f_{r+1} + \ldots + f_d)(x_k)|$.

Our intention is to show that $|f(x_k)| \leq \frac{2}{m^2_{2j+1}}$ for $k > r$. The same argument also gives that $|f(x_r)| \leq \frac{20}{m_{2j+1}} + \frac{2}{m^2_{2j+1}} < \frac{21}{m_{2j+1}}$.

Let $k > r$. By our assumptions and the definition of the injection σ we get that $w(f_l) \neq m_{2j_k}$ for $l = r, r+1, \ldots d$. We set:

$$K_1 = \{l : \ w(f_l) < m_{2j_k}\} \quad \text{and} \quad K_2 = \{l : \ w(f_l) > m_{2j_k}\}$$

As we have already seen we have $|f_l(x_k)| \leq \frac{28}{w(f_l)}$ for all $l \in K_1$. Thus

$$\sum_{l \in K_1} |f_l(x_k)| \leq \sum_{l \in K_1} \frac{28}{w(f_l)} < \frac{1}{m_{2j+1}}. \tag{17}$$

We also set

$$F^1_k = \Big\{ i \in F_k : \ \#\{l \in K_2 : \ \operatorname{ran} x^k_i \cap \operatorname{ran} f_l \neq \emptyset\} \leq 1 \Big\} \quad \text{and} \quad F^2_k = F_k \setminus F^1_k.$$

The set $\{\min\operatorname{supp} x^k_i \ : \ i \in F^2_k\}$ belongs to $\mathcal{S}_{n_{2j+1}+1}$ thus, since $\sum_{i \in F_k} a^k_i x^k_i$ is a $(\frac{1}{4m^2_{2j_k}}, 2j_k)$ S.C.C. we get that $\sum_{i \in F^2_k} a^k_i < \frac{1}{4m^2_{2j_k}}$. It follows that

$$|(\sum_{l \in K_2} f_l)(\sum_{i \in F^2_k} a^k_i x^k_i)| \leq m_{2j+1} \sum_{i \in F^2_k} a^k_i \|x^k_i\| \leq m_{2j+1} \cdot \frac{1}{4m^2_{2j_k}} \cdot 4 = \frac{m_{2j+1}}{m^2_{2j_k}}. \tag{18}$$

Since the sequence $(x^k_i)_i$ is a $(4, \frac{1}{m^2_{2j_k}})$ R.I.S. for each $l \in K_2$ we may select by the basic inequality (Proposition 4.3) an $h^1_l \in \operatorname{conv}_{\mathbb{Q}}\{h \in B : \ w(h) = w(f_l)\}$ a $t_l \in \mathbb{N}$ with $t_l \notin \operatorname{supp} h^1_l$ and an h^2_l with $\|h^2_l\|_\infty \leq \frac{1}{m^2_{2j_k}}$ such that the functionals $h^1_l + e^*_{t_l}$ and h^2_l are supported on $\{t^k_i \ : \ i \in F^1_k \text{ and } \operatorname{ran} x^k_i \cap \operatorname{ran} f_l \neq \emptyset\}$ (where $t^k_i = \min\operatorname{supp} x^k_i$ for all i) and

$$|f_l(\sum_{i \in F^1_k} a^k_i x^k_i)| \leq 4(2(h^1_l + e^*_{t_l}) + h^2_l)(\sum_{i \in F^1_k} a^k_i e_{t^k_i}). \tag{19}$$

Notice that, as follows from the definition of the set F^1_k, the functionals $(h^1_l + e^*_{t_l} + h^2_l)_{l \in K_2}$ have pairwise disjoint supports. The set $\{t_l : \ l \in K_2\}$ belongs to $\mathcal{S}_{n_{2j+1}+1}$

thus $(\sum_{l\in K_2} e^*_{t_l})(\sum_{i\in F^1_k} a^k_i e_{t^k_i}) < \frac{1}{4m^2_{2j_k}}$. It follows from (19) that

$$\begin{aligned}\sum_{l\in K_2} |f_l(\sum_{i\in F^1_k} a^k_i x^k_i)| &\leq (8\sum_{l\in K_2} h^1_l + 8\sum_{l\in K_2} e^*_{t_l} + 4\sum_{l\in K_2} h^2_l)(\sum_{i\in F^1_k} a^k_i e_{t^k_i}) \\ &\leq 8\sum_{l\in K_2}\frac{1}{w(f_l)} + 8\cdot\frac{1}{4m^2_{2j_k}} + 4\cdot\frac{1}{m^2_{2j_k}} \leq \frac{7}{m^2_{2j_k}}. \qquad (20)\end{aligned}$$

Inequalities (17),(18) and (20) yield that for $k > r$ we have

$$\begin{aligned}|f(x_k)| \leq \frac{1}{m_{2j+1}} &\Big(|\sum_{l\in K_1} f_l(x_k)| + m_{2j_k}|\sum_{l\in K_2} f_l(\sum_{i\in F^1_k} a^k_i x^k_i)| \\ &+ m_{2j_k}\sum_{l\in K_2}|f_l(\sum_{i\in F^2_k} a^k_i x^k_i)|\Big) \\ &\leq \frac{1}{m_{2j+1}}\Big(\frac{1}{m_{2j+1}} + m_{2j_k}\frac{m_{2j+1}}{m^2_{2j_k}} + m_{2j_k}\frac{7}{m^2_{2j_k}}\Big) \leq \frac{2}{m^2_{2j+1}}.\end{aligned}$$

Let now E be an interval of $\{1,2,\dots,n\}$. Set $E_1 = E\cap\{t-1\}$, $E_2 = E\cap\{t,t+1,\dots,r-1\}$, $E_3 = E\cap\{r\}$ and $E_4 = E\cap\{r+1,\dots,d\}$. We recall that the sequence $(a_k)_{k=1}^n$ is decreasing and thus we have that

$$\begin{aligned}|f(\sum_{k\in E}(-1)^k a_k x_k)| &\leq \sum_{i=1}^4 |f(\sum_{k\in E_i}(-1)^k a_k x_k)| \\ &\leq \frac{20}{m_{2j+1}}\sum_{k\in E_1} a_k + \frac{\theta}{m_{2j+1}}|\sum_{k\in E_2}(-1)^k a_k| + \frac{21}{m_{2j+1}}\sum_{k\in E_3} a_k + \frac{2}{m^2_{2j+1}}\sum_{k\in E_4} a_k \\ &\leq \frac{20}{m_{2j+1}}\max_{k\in E} a_k + \frac{\theta}{m_{2j+1}}\max_{k\in E} a_k + \frac{21}{m_{2j+1}}\max_{k\in E} a_k + \frac{2}{m^2_{2j+1}}\sum_{k\in E} a_k \\ &\leq \max_{k\in E} a_k + 2\varepsilon\sum_{k\in E} a_k < 28(2\max_{k\in E} a_k + \varepsilon\sum_{k\in E} a_k)\end{aligned}$$

So assumption (d) of the basic inequality (Proposition 4.3) is fulfilled for the linear combination $\sum_{k=1}^n (-1)^k a_k x_k$.

The proof of the proposition is complete. □

Proof of Theorem 7.3. It is enough to show that for every pair Y, Z of block subspaces of X_G and $\delta > 0$ there exist $y\in Y$ and $z \in Z$ such that $\|y-z\| < \delta\|y+z\|$.

Let such Y, Z, δ and select $j\in\mathbb{N}$ with $m_{2j+1} > \frac{280}{\delta}$. Our goal is to find a $(\frac{1}{2}, 2j+1)$ dependent couple $(x_k, x^*_k)_{k\in\mathbb{N}}$ satisfying the additional property that $x_k \in Y$ if k is odd while $x_k \in Z$ if k is even. When such a choice of the dependent couple is done we consider the vector $a(n_{2j+1}, (x_k)_{k\in\mathbb{N}}) = \sum_{k=1}^n a_k x_k$. Then $(x^*_k)_{k=1}^n$ is an $\mathcal{S}_{n_{2j+1}}$ admissible special sequence hence the functional

$$f = \frac{1}{m_{2j+1}}(x^*_1 + x^*_2 + \cdots + x^*_n)$$

belongs to the norming set D. Thus

$$\|\sum_{k=1}^{n} a_k x_k\| \geq f(\sum_{k=1}^{n} a_k x_k) = \frac{1}{2m_{2j+1}}. \tag{21}$$

On the other hand Proposition 7.5 yields that

$$\|\sum_{k=1}^{n} (-1)^k a_k x_k\| \leq \frac{140}{m_{2j+1}^2}. \tag{22}$$

We set

$$y = \sum_{k\leq n,\ k \text{ odd}} a_k x_k \text{ and } z = \sum_{k\leq n,\ k \text{ even}} a_k x_k.$$

Clearly $y \in Y$ and $z \in Z$. By (21),(22) and the fact that $m_{2j+1} > \frac{280}{\delta}$ we get that

$$\|y - z\| < \delta\|y + z\|.$$

We proceed to the selection of a $(\frac{1}{2}, 2j+1)$ dependent couple $(x_k, x_k^*)_{k\in\mathbb{N}}$ with $x_{2k-1} \in Y$ and $x_{2k} \in Z$.

In the first step x_1 and x_1^* are defined. Choose $j_1 \geq j + 2$. Recall that the norming set D is $2l$ closed for all $l \in \mathbb{N}$. We thus may select, with consecutive applications of Proposition 5.4, a block sequence $(y_i)_{i\in\mathbb{N}}$ in Y and a strictly increasing sequence of integers $(l_i)_{i\in\mathbb{N}}$ with $\min\operatorname{supp} y_1 > 3m_{2j+1}^2$ such that each y_i is a $(\frac{1}{m_{2l_i}}, 2l_i)$ seminormalized S.C.C. Lemma 6.5 yields a subsequence $(y_{k_i})_{i\in\mathbb{N}}$ of $(y_i)_{i\in\mathbb{N}}$ such that its sequence of successive differences, i.e. the sequence $(x_i^1)_{i\in\mathbb{N}}$ defined by $x_i^1 = y_{k_{2i-1}} - y_{k_{2i}}$, is a $(4, \frac{1}{m_{2j_1}^2})$ R.I.S. Passing to a further subsequence we may assume that the vector $a(n_{2j_1}, (x_i^1)_i) = \sum_{i\in F_1} a_i^1 x_i^1$ is a $(\frac{1}{4m_{2j_1}^2}, 2j_1)$ S.C.C. By Proposition 6.2(1a) we get that $\|\sum_{i\in F_1} a_i^1 x_i^1\| \leq \frac{20}{m_{2j_1}}$. For each $i \in F_1$ select $f_i^1 \in D$ with $\operatorname{ran} f_i^1 = \operatorname{ran} x_i^1$ such that $f_i^1(x_i^1) \geq \frac{1}{2}$. We set

$$x_1^* = \frac{1}{m_{2j_1}} \sum_{i\in F_1} f_i^1.$$

The functional x_1^* belongs to $A \cap D$ and $w(f_1) = m_{2j_1}$. We thus have $\|\sum_{i\in F_1} a_i^1 x_i^1\| \geq x_1^*(\sum_{i\in F_1} a_i^1 x_i^1) \geq \frac{1}{2m_{2j_1}}$. We choose $\frac{1}{40} \leq \theta_1 \leq 1$ such that $x_1^*(\theta_1 m_{2j_1} \sum_{i\in F_1} a_i^1 x_i^1) = \frac{1}{2}$ and we set

$$x_1 = \theta_1 m_{2j_1} \sum_{i\in F_1} a_i^1 x_i^1.$$

Observe that $x_1 \in Y$, $\operatorname{ran} x_1 = \operatorname{ran} x_1^*$ and $x_1^*(x_1) = \frac{1}{2}$; the first step of the choice of the dependent couple is complete.

We describe briefly the next steps. Let $\sigma(x_1^*) = 2j_2$ and apply the same procedure as before to find x_2^* with $w(x_2^*) = m_{2j_2}$ and a vector $x_2 = \theta_2 m_{2j_2} \sum_{i\in F_2} a_i^2 x_i^2$ in Z satisfying the desired properties of a dependent couple and additionally $x_1 < x_2$. Then $\sigma(x_1^*, x_2^*) = 2j_3$ and we continue as before.

The proof of Theorem 7.3 is complete. □

CHAPTER 8

The predual $(X_G)_*$ of X_G is also H.I.

Our next task is to show that, under the assumption that G is $\mathcal{S}_2$ bounded, the space

$$(X_G)_* = \overline{\text{span}}\{e_n^* : n \in \mathbb{N}\}$$

which is the predual of the space $X_G = T[G, (\mathcal{S}_{n_j}, \frac{1}{m_j})_j, D]$, is a H.I. space. It is worthwhile to point out that this property is not true in general for H.I. Banach spaces. For example there are reflexive H.I. spaces X such that X^* contains ℓ^2 or any ℓ^p, $1 < p < \infty$ ([**9**]). From a technical point of view the proof of the H.I. property of $(X_G)_*$ lies on the rational convexity of the set D. The same property is also used in the study of the space $\mathcal{L}(X_G)$.

LEMMA 8.1. *Let $h \in \text{span}\{e_n^* : n \in \mathbb{N}\}$ with $\|h\| \leq 1$. Then for every $\delta > 0$ there exists an $f \in D$ such that $\text{ran}\, f = \text{ran}\, h$ and $\|f - h\| < \delta$.*

Proof. Since $B_{X_G^*} = \overline{D}^p$ we can find an $f \in D$ such that $\text{ran}\, f = \text{ran}\, h$ and $|f(e_k) - h(e_k)| < \dfrac{\delta}{\#(\text{ran}\, h)}$ for each $k \in \text{ran}\, h$. It follows from the triangle inequality that f satisfies the desired property. □

LEMMA 8.2. *Let Z be a block subspace of the basis $(e_k^*)_{k\in\mathbb{N}}$ of $(X_G)_*$, $\delta > 0$ and $j \in \mathbb{N}$ with $j > 1$. Then there exists a vector $x \in X_G$ and a functional $h \in D$ such that*

(i) $\text{dist}(h, Z) < \delta$.

(ii) *The vector x is a $(\frac{1}{m_{2j}}, 2j)$ S.C.C. of a normalized block sequence.*

(iii) $h(x) > \frac{1}{3}$ *and* $\text{ran}\, h = \text{ran}\, x$.

Proof. We may assume that $\delta < \frac{1}{3}$. Let $(z_k^*)_{k\in\mathbb{N}}$ be a normalized block sequence in Z and choose, using Lemma 8.1, a block sequence $(f_k)_{k\in\mathbb{N}}$ in D with $\text{ran}\, f_k = \text{ran}\, z_k^*$ and $\sum_{k=1}^{\infty} \|z_k^* - f_k\| < \frac{\delta}{2}$.

Proposition 5.5 yields that there exists a finite block sequence $(h_k)_{k\in F}$ of $(f_k)_{k\in\mathbb{N}}$ such that

(1) $\|h_k\| > \frac{1}{2}$ for all k and $\|\sum_{k\in F} h_k\| \leq 1$.

(2) The set $\{\min\text{supp}\, h_k : k \in F\}$ is a maximal $\mathcal{S}_{2\log_2(m_{2j})(n_{2j-1}+1)+1}$ set.

(3) There exists a $c < 1$ such that each h_k is of the form $h_k = c \sum_{i\in F_k} f_i$.

For each $k \in F$ select a $x_k \in X_G$ with $\text{ran}\, x_k = \text{ran}\, h_k$, $\|x_k\| = 1$ and $h_k(x_k) > \frac{1}{2}$. Since $\|\sum_{k\in F} h_k\| \leq 1$ we may select an $h \in D$ with $\text{ran}\, h = \text{ran} \sum_{k\in F} h_k$ such that

$\|h - \sum_{k\in F} h_k\| < \frac{\delta}{2}$. We have that

$$\begin{aligned}\operatorname{dist}(h, Z) &\leq \|h - \sum_{k\in F} h_k\| + \operatorname{dist}(\sum_{k\in F} h_k, Z)\\ &< \frac{\delta}{2} + c\sum_{k=1}^{\infty} \|z_k^* - f_k\| < \delta\end{aligned}$$

We consider the vector

$$x = a\Big(2\log_2(m_{2j})(n_{2j-1}+1)+1, (x_k)_{k\in\mathbb{N}}\Big) = \sum_{k=1}^{n} a_k x_k$$

Obviously x is a $(\frac{1}{m_{2j}}, 2j)$ S.C.C. of $(x_k)_{k\in\mathbb{N}}$ and $\operatorname{ran} h = \operatorname{ran} x$.

For $k = 1, 2, \dots$ we have that $h(x_k) > h_k(x_k) - \|h - \sum_{k\in F} h_k\| > \frac{1}{2} - \frac{\delta}{2} > \frac{1}{3}$. Therefore

$$h(x) = \sum_{k=1}^{n} a_k h(x_k) > \frac{1}{3}.$$

□

THEOREM 8.3. *If the ground set G is $\mathcal{S}_2$ bounded then the predual $(X_G)_* = \overline{\operatorname{span}}\{e_n^* : n \in \mathbb{N}\}$ of the space $X_G = T[G, (\mathcal{S}_{n_j}, \frac{1}{m_j})_j, D]$ is H.I.*

Proof. Let Y, Z be a pair of block subspaces of $(X_G)_*$ and $\delta > 0$. We will find $h_Y \in Y$ and $h_Z \in Z$ such that $\|h_Y - h_Z\| < \delta\|h_Y + h_Z\|$.

Let such Y, Z, δ and choose $j \in \mathbb{N}$ such that $m_{2j+1} > 420(\frac{2}{\delta} + 1)$. Our goal is to create a $(\frac{1}{3}, 2j+1)$ dependent couple $(x_k, x_k^*)_{k\in\mathbb{N}}$ satisfying additionally $\operatorname{dist}(x_k^*, Y) < \frac{1}{2^k}$ if k is odd while $\operatorname{dist}(x_k^*, Z) < \frac{1}{2^k}$ if k is even. We first describe how we finish the proof when the existence of such a dependent couple is verified.

We consider the vector $a(n_{2j+1}, (x_k)_{k\in\mathbb{N}}) = \sum_{k=1}^{n} a_k x_k$. The family $(x_k^*)_{k=1}^{n}$ is an $\mathcal{S}_{n_{2j+1}}$ admissible special sequence thus the functional

$$f = \frac{1}{m_{2j+1}}(x_1^* + x_2^* + \cdots + x_n^*)$$

is a member of D hence

$$\|f\| \leq 1 \tag{23}$$

$$\text{and} \quad \|\sum_{k=1}^{n} a_k x_k\| \geq f(\sum_{k=1}^{n} a_k x_k) = \frac{1}{3m_{2j+1}}. \tag{24}$$

By Proposition 7.5 we also have that

$$\|\sum_{k=1}^{n}(-1)^k a_k x_k\| \leq \frac{140}{m_{2j+1}^2}. \tag{25}$$

We set

$$f' = \frac{1}{m_{2j+1}} \sum_{k=1}^{n} (-1)^k x_k^* \tag{26}$$

and from (24) and (25) we get that

$$\|f'\| \geq \frac{f'\Big(\sum\limits_{k=1}^{n}(-1)^k a_k x_k\Big)}{\|\sum\limits_{k=1}^{n}(-1)^k a_k x_k\|} \geq \frac{\frac{1}{3m_{2j+1}}}{\frac{140}{m_{2j+1}^2}} = \frac{m_{2j+1}}{420}. \tag{27}$$

We set

$$f_Y = \frac{1}{m_{2j+1}} \sum_{k\leq n,\ k \text{ odd}} x_k^* \quad \text{and} \quad f_Z = -\frac{1}{m_{2j+1}} \sum_{k\leq n,\ k \text{ even}} x_k^*.$$

We have that

$$\operatorname{dist}(f_Y, Y) \leq \frac{1}{m_{2j+1}} \sum_{k \text{ odd}} \operatorname{dist}(x_k^*, Y) < \frac{1}{m_{2j+1}} < \frac{1}{2} \quad \text{and} \quad \operatorname{dist}(f_Z, Z) < \frac{1}{2}$$

thus we may choose $h_Y \in Y$ and $h_Z \in Z$ such that

$$\|f_Y - h_Y\| < \frac{1}{2} \quad \text{and} \quad \|f_Z - h_Z\| < \frac{1}{2}. \tag{28}$$

From (23) and (28) it follows that

$$\|h_Y - h_Z\| \leq \|h_Y - f_Y\| + \|f_Y - f_Z\| + \|f_Z - h_Z\| < \frac{1}{2} + \|f\| + \frac{1}{2} \leq 2 \tag{29}$$

while from (27) and (28) we get that

$$\|h_Y + h_Z\| \geq \|f_Y + f_Z\| - \|h_Y - f_Y\| - \|h_Z - f_Z\| > \|f'\| - \frac{1}{2} - \frac{1}{2} > \frac{m_{2j+1}}{420} - 1. \tag{30}$$

Therefore (29),(30) and the fact that $m_{2j+1} > 420(\frac{2}{\delta} + 1)$ yield that

$$\|h_Y - h_Z\| < \delta \|h_Y + h_Z\|.$$

We next proceed to the inductive construction of a $(\frac{1}{3}, 2j+1)$ dependent couple $(x_k, x_k^*)_{k\in\mathbb{N}}$ with $\operatorname{dist}(x_{2k-1}^*, Y) < \frac{1}{2^{2k-1}}$ and $\operatorname{dist}(x_{2k}^*, Z) < \frac{1}{2^{2k}}$. Let $k \in \mathbb{N}$. If $k = 1$ we select $j_1 \geq j + 2$ while for $k \geq 2$, if $(x_i, x_i^*)_{i<k}$ have been already defined, we set $2j_k = \sigma(x_1^*, \ldots, x_{k-1}^*)$. Our intention is to define x_k, x_k^* satisfying the properties of a $(\frac{1}{3}, 2j+1)$ dependent couple and moreover $\operatorname{dist}(x_k^*, Y) < \frac{1}{2^k}$ if k is odd while $\operatorname{dist}(x_k^*, Z) < \frac{1}{2^k}$ if k is even. Suppose that k is even. We may select, by consecutive applications of Lemma 8.2, a block sequence $(z_i)_{i\in\mathbb{N}}$ in X_G with $\min\operatorname{supp} z_1 > \max\operatorname{supp} x_{k-1}$ (or $\min\operatorname{supp} z_1 > 3m_{2j+1}^2$ if $k = 1$), a strictly increasing sequence of integers $(l_i)_{i\in\mathbb{N}}$ and a sequence $(f_i)_{i\in\mathbb{N}}$ in D such that

(1) $\operatorname{dist}(f_i, Z) < \frac{1}{2^i}$.

(2) The vector z_i is a $(\frac{1}{m_{2l_i}}, 2l_i)$ S.C.C. of a normalized block sequence.

(3) $f_i(z_i) > \frac{1}{3}$ and $\operatorname{ran} f_i = \operatorname{ran} z_i$.

Lemma 6.5 yields a subsequence $(z_{t_i^k})_{i\in\mathbb{N}}$ of $(z_i)_{i\in\mathbb{N}}$ such that its sequence of successive differences, i.e. the sequence $(x_i^k)_{i\in\mathbb{N}}$ defined by $x_i^k = z_{t_{2i-1}^k} - z_{t_{2i}^k}$, is a $(4, \frac{1}{m_{2j_k}^2})$ R.I.S. Passing to a further subsequence we may assume that the vector $a(n_{2j_k}, (x_i^k)_i) = \sum_{i\in F_k} a_i^k x_i^k$ is a $(\frac{1}{4m_{2j_k}^2}, 2j_k)$ S.C.C. We set

$$x_k^* = \frac{1}{m_{2j_k}}\Big(\sum_{i\in F_k\setminus\{\max F_k\}} f_{t_{2i-1}^k} - f_{t_{2\max F_k}^k}\Big).$$

The functional x_k^* is a member of $A\cap D$ with $w(x_k^*) = m_{2j_k}$. We get that $\|\sum_{i\in F_k} a_i^k x_i^k\| \geq x_k^*(\sum_{i\in F_k} a_i^k x_i^k) > \frac{1}{3m_{2j_k}}$. On the other hand by Proposition 6.2(1a) we get that $\|\sum_{i\in F_k} a_i^k x_i^k\| \leq \frac{20}{m_{2j_k}}$. We select $\frac{1}{60} \leq \theta_k < 1$ such that $x_k^*(\theta_k m_{2j_k} \sum_{i\in F_k} a_i^k x_i^k) = \frac{1}{3}$ and we set

$$x_k = \theta_k m_{2j_k} \sum_{i\in F_k} a_i^k x_i^k.$$

Observe that $x_k^*(x_k) = \frac{1}{3}$, $\operatorname{ran} x_k = \operatorname{ran} x_k^*$ and

$$\operatorname{dist}(x_k^*, Z) < \frac{1}{m_{2j_k}} \sum_{i=1}^{\infty} \operatorname{dist}(f_i, Z) < \frac{1}{2^k}.$$

If k is odd the only difference in the previous construction is that x_k^* is selected to satisfy $\operatorname{dist}(x_k^*, Y) < \frac{1}{2^k}$.

The inductive construction of the $(\frac{1}{3}, 2j+1)$ dependent couple is complete and this finishes the proof of Theorem 8.3. □

CHAPTER 9

The structure of the space of operators $\mathcal{L}(X_G)$

We next show that the assumption that the ground set G is $\mathcal{S}_2$ bounded yields that every bounded linear operator acting on X_G, is of the form $T = \lambda I + S$ with $\lambda \in \mathbb{R}$ and S strictly singular. For the proof of this result we use ideas from the proof of the analogous results in [**28**] (Lemmas 22,23). We also show that every strictly singular operator is weakly compact and moreover that the space X_G is somewhat reflexive.

LEMMA 9.1. Suppose that the ground set G is $\mathcal{S}_2$ bounded. Let $(y_k)_{k\in\mathbb{N}}$ be a bounded block sequence in X_G and Y a closed subspace of X_G containing the sequence $(y_k)_{k\in\mathbb{N}}$. Let also $T : Y \longrightarrow X_G$ be a bounded linear operator such that the following conditions are fulfilled.

(1) There exists a strictly increasing sequence $(i_k)_{k\in\mathbb{N}}$ of integers such that each y_k is a seminormalized $(\frac{1}{m_{2i_k}}, 2i_k)$ S.C.C.

(2) $(Ty_k)_{k\in\mathbb{N}}$ is a block sequence in X_G.

Then

$$\operatorname{dist}(Ty_n, \mathbb{R}y_n) \overset{n\to\infty}{\longrightarrow} 0.$$

Proof. Assume the contrary. Then $\operatorname{dist}(Ty_n, \mathbb{R}y_n) > \delta$ for infinitely many n's, for some $\delta > 0$. Furthermore we may assume, passing to a subsequence, that $\operatorname{dist}(Ty_n, \mathbb{R}y_n) > \delta$ and $\operatorname{supp}(y_n) \cup \operatorname{supp}(Ty_n) < \operatorname{supp}(y_{n+1}) \cup \operatorname{supp}(Ty_{n+1})$ for all n. By standard perturbation arguments we may also assume that $\operatorname{ran} y_n = \operatorname{ran} Ty_n$. Hahn-Banach Theorem yields that there exists $y_n^* \in \mathrm{B}_{X_G^*}$ such that $y_n^*(Ty_n) > \delta$ and $y_n^*(y_n) = 0$. Since $\mathrm{B}_{X_G^*} = \overline{D}^p$ and from the fact that the basis is bimonotone we may assume that $y_n^* \in D$ and $\operatorname{ran} y_n^* = \operatorname{ran} Ty_n = \operatorname{ran} y_n$.

We choose $j \in \mathbb{N}$ with

$$m_{2j+1} > \frac{140 \cdot \|T\|}{\delta}$$

Our goal is to find a $(0, 2j+1)$ dependent couple $(x_k, x_k^*)_{k\in\mathbb{N}}$ with $x_k \in Y$ for all k satisfying $x_k^*(Tx_k) < -\delta$ if k is odd and $x_k^*(Tx_k) > \delta$ if k is even. We first show how the existence of such a dependent couple leads to a contradiction.

We consider the vector $a(n_{2j+1}, (x_k)_{k\in\mathbb{N}}) = \sum\limits_{k=1}^{n} a_k x_k$ and we set

$$x = \sum_{k=1}^{n} (-1)^k a_k x_k.$$

Then $\{x_k^* : k = 1, 2, \ldots, n\}$ is a special $\mathcal{S}_{n_{2j+1}}$ admissible sequence with $w(x_1^*) > m_{2j+1}$ thus the functional $x^* = \dfrac{1}{m_{2j+1}}(x_1^* + \cdots + x_n^*)$ belongs to the norming set

D therefore

$$\|Tx\| \geq x^*(Tx) = \frac{1}{m_{2j+1}} \sum_{k=1}^{n} a_k(-1)^k x_k^*(Tx_k) > \frac{\delta}{m_{2j+1}}. \tag{31}$$

On the other hand Proposition 7.5 yields that

$$\|x\| = \|\sum_{k=1}^{n}(-1)^k a_k x_k\| \leq \frac{140}{m_{2j+1}^2}. \tag{32}$$

Inequalities (31) and (32) yield that $\|T\| > \dfrac{\delta \cdot m_{2j+1}}{140}$ which contradicts to our choice of j.

We proceed with the construction of the desired $(0, 2j+1)$ dependent couple $(x_k, x_k^*)_{k\in\mathbb{N}}$ with $x_k \in Y$ for all k and $x_k^*(Tx_k) < -\delta$ for k odd, $x_k^*(Tx_k) > \delta$ for k even.

Let $k \in \mathbb{N}$ and suppose that $(x_i, x_i^*)_{i<k}$ have been defined satisfying the inductive assumptions. If $k = 1$ we select $j_1 \geq j+2$ while if $k > 1$ we set $2j_k = \sigma(x_1^*, \ldots, x_{k-1}^*)$. By Lemma 6.5 we may select a subsequence $(y_{t_i^k})_{i\in\mathbb{N}}$ of $(y_i)_{i\in\mathbb{N}}$ with $\min\operatorname{supp} y_{t_1^k} > \max\operatorname{supp} x_{k-1}$ ($\min\operatorname{supp} y_{t_1^k} > 3m_{2j+1}^2$ if $k=1$) such that the sequence $(x_i^k)_{i\in\mathbb{N}}$ defined by $x_i^k = y_{t_{2i-1}^k} - y_{t_{2i}^k}$ is a $(4, \frac{1}{m_{2j_k}^2})$ R.I.S. Passing to a further subsequence we may assume that the vector $a(n_{2j_k}, (x_i^k)_i) = \sum_{i\in F_k} a_i^k x_i^k$ is a $(\frac{1}{4m_{2j_k}^2}, 2j_k)$ S.C.C. We set

$$x_k = m_{2j_k} \sum_{i\in F_k} a_i^k x_i^k \quad \text{and} \quad x_k^* = \frac{(-1)^k}{m_{2j_k}} \Big(\sum_{i\in F_k\setminus\{\max F_k\}} y_{t_{2i-1}^k}^* - y_{t_{2\max F_k}^k}^* \Big).$$

We have that $x_k \in Y$, $x_k^* \in A \cap D$ with $w(x_k^*) = m_{2j_k}$, $\operatorname{ran} x_k = \operatorname{ran} x_k^*$, $x_k^*(x_k) = 0$ and $x_k^*(Tx_k) < -\delta$ if k is odd while $x_k^*(Tx_k) > \delta$ if k is even. The inductive construction of the $(0, 2j+1)$ dependent couple $(x_k, x_k^*)_{k\in\mathbb{N}}$ is complete. This finishes the proof of Lemma 9.1. □

THEOREM 9.2. *Let G be a ground set which is $\mathcal{S}_2$ bounded. Then every bounded linear operator in $\mathcal{L}(X_G)$ is of the form $T = \lambda I + S$ where $\lambda \in \mathbb{R}$ and S is a strictly singular operator.*

Proof. Suppose that T is not strictly singular. We shall determine a $\lambda \in \mathbb{R}$, such that $T - \lambda I$ is a strictly singular operator. Let Y be an infinite dimensional subspace of X_G such that the restriction of T to Y is an isomorphic embedding. Using standard perturbation arguments and passing to a further subspace we may assume that there exists a normalized block sequence $(z_n)_{n\in\mathbb{N}}$, such that $(Tz_n)_{n\in\mathbb{N}}$ is a block sequence and $Y = \overline{\operatorname{span}}\{z_n : n \in \mathbb{N}\}$.

Select a block sequence $(y_k)_{k\in\mathbb{N}}$ in Y and a strictly increasing sequence $(i_k)_{k\in\mathbb{N}}$ of integers such that each y_k is a seminormalized $(\frac{1}{m_{2i_k}}, 2i_k)$ S.C.C. Since by our assumptions $(Ty_k)_{k\in\mathbb{N}}$ is also a block sequence Lemma 9.1 implies that

$$\operatorname{dist}(Ty_n, \mathbb{R}y_n) \overset{n\to\infty}{\longrightarrow} 0$$

Let $(\lambda_n)_{n\in\mathbb{N}}$ be a sequence of reals with

$$\|Ty_n - \lambda_n y_n\| \overset{n\to\infty}{\longrightarrow} 0.$$

Since T is a bounded operator and $\|y_n\| \geq \frac{1}{2}$ we get that $(\lambda_n)_{n\in\mathbb{N}}$ is a bounded sequence so it has a subsequence converging to some $\lambda \in \mathbb{R}$. We may assume that $\lambda_n \longrightarrow \lambda$.

We set $\Lambda = \overline{\text{span}}\{y_n : n \in \mathbb{N}\}$. It is obvious that the restriction of the operator $T - \lambda I$ to any finite codimensional subspace of Λ is not an isomorphism. Hence, Proposition 1.2 yields that the operator $T - \lambda I$ is strictly singular. □

The following lemma is slight modification of Lemma 9.1 and its proof follows the same lines as that of Lemma 9.1 so we omit it.

LEMMA 9.3. Suppose that the ground set G is $\mathcal{S}_2$ bounded. Let $(y_k)_{k\in\mathbb{N}}$ be a bounded block sequence in X_G and Y a closed subspace of X_G containing the sequence $(y_k)_{k\in\mathbb{N}}$. Let also $T : Y \longrightarrow X_G$ be a bounded linear operator such that the following conditions are fulfilled:

(1) There exists a strictly increasing sequence $(i_k)_{k\in\mathbb{N}}$ of integers such that each y_k is a $(\frac{1}{m_{2i_k}}, 2i_k)$ S.C.C. of a normalized sequence and $\|y_k\| > \delta > 0$ for all k.
(2) The sequence $(w_k)_{k\in\mathbb{N}}$ defined by $w_k = y_{2k-1} - y_{2k}$ satisfies the property that $(Tw_k)_{k\in\mathbb{N}}$ is a block sequence in X_G.

Then

$$\text{dist}(Tw_n, \mathbb{R}w_n) \overset{n\to\infty}{\longrightarrow} 0.$$

THEOREM 9.4. *Let G be a ground set which is $\mathcal{S}_2$ bounded. Then every bounded linear operator in $\mathcal{L}(X_G)$ is of the form $T = \lambda I + S$ where $\lambda \in \mathbb{R}$ and S is a weakly compact operator.*

Proof. By Theorem 9.2 it is enough to show that if T is a non weakly compact operator on X_G then T is not strictly singular. Let $T : X_G \longrightarrow X_G$ be a non weakly compact bounded linear operator. Then there exists a normalized sequence $(x_n)_{n\in\mathbb{N}}$ in X_G such that the sequence $(Tx_n)_{n\in\mathbb{N}}$ has no weakly convergent subsequence. Since X_G contains no isomorphic copy of ℓ^1 we may assume, passing to a subsequence, that $(Tx_n)_{n\in\mathbb{N}}$ is nontrivial weakly Cauchy. We set y^{**} the w^* limit of $(Tx_n)_{n\in\mathbb{N}}$. We may also assume that there exists a $x^{**} \in X_G^{**} \setminus X_G$ such that x^{**} is the w^* limit of $(x_n)_{n\in\mathbb{N}}$. Obviously $y^{**} = T^{**}x^{**}$.

Since the Schauder basis $(e_n)_{n\in\mathbb{N}}$ of X_G is boundedly complete, using standard perturbation arguments, we may assume that $x_n = x + u_n$ where $x = \sum_{i=1}^{\infty} x^{**}(e_i^*)e_i$ and $(u_n)_{n\in\mathbb{N}}$ is a block sequence. Clearly $u_n \overset{w^*}{\longrightarrow} x^{**} - x$ and also $Tu_n \overset{w^*}{\longrightarrow} y^{**} - Tx \in X_G^{**} \setminus X_G$. Therefore we may assume that $x = 0$ which yields that $(x_n)_{n\in\mathbb{N}}$ is a block sequence. Using similar perturbation arguments we may assume that for each n, $Tx_n = z + z_n$ where $z \in X_G$ and $(z_n)_{n\in\mathbb{N}}$ is a block sequence. Set

$$\theta = \text{dist}(y^{**}, X_G).$$

Since $z \in X_G$, $z = \sum_{n=1}^{\infty} a_n e_n$ and there exists $i_0 \in \mathbb{N}$ such that

$$\|\sum_{i=i_0+1}^{\infty} a_i e_i\| < \frac{\theta}{4}. \tag{33}$$

Select a $y^* \in \mathrm{B}_{X_G^*}$ such that $y^{**}(y^*) > \frac{3\theta}{4}$ and y^* restricted to $\mathrm{span}\{e_1, e_2, \ldots, e_{i_0}\}$ is zero. The sequence $(y^*(Tx_n))_{n\in\mathbb{N}}$ tends to $y^{**}(y^*)$ so we may assume that $y^*(Tx_n) = y^*(z + z_n) > \frac{3\theta}{4}$ for all n. Relation (33) also yields that $|y^*(z)| < \frac{\theta}{4}$ thus

$$y^*(z_n) > \frac{\theta}{2} \qquad \text{for all } n.$$

Let $x^* \in \mathrm{B}_{X_G^*}$ and $\delta > 0$ with $x^{**}(x^*) > \delta$. Since x^{**} is the w^* limit of the sequence $(x_n)_{n\in\mathbb{N}}$ we may assume that $x^*(x_n) > \delta$ for all n. We select a block sequence $(y_k)_{k\in\mathbb{N}}$ of $(x_n)_{n\in\mathbb{N}}$ and a strictly increasing sequence $(i_k)_{k\in\mathbb{N}}$ such that each y_k is an $(\frac{1}{m_{2i_k}}, 2i_k)$ S.C.C. of the sequence $(x_n)_{n\in\mathbb{N}}$. The facts that $x^*(x_n) > \delta$ and $y^*(Tx_n) > \frac{3\theta}{4}$ for all n in conjunction to the fact that each y_k is a convex combination of the sequence $(x_n)_{n\in\mathbb{N}}$ yield that $\|y_k\| \geq x^*(y_k) > \delta$ and $y^*(Ty_k) > \frac{3\theta}{4}$ for all k. Since also each Ty_k is a convex combination of the sequence $(z + z_n)_{n\in\mathbb{N}}$ there exists a block sequence $(v_k)_{k\in\mathbb{N}}$ such that for each k the vector v_k is a convex combination of $(z_n)_{n\in\mathbb{N}}$ and $Ty_k = z + v_k$. We thus get that $\|v_k\| \geq y^*(Ty_k) - y^*(z) > \frac{3\theta}{4} - \frac{\theta}{4} = \frac{\theta}{2}$ for all k. Observe that $(Tw_k)_{k\in\mathbb{N}}$ is a block sequence where $(w_k)_{k\in\mathbb{N}}$ is the sequence defined by $w_k = y_{2k-1} - y_{2k}$. Lemma 9.3 yields that

$$\mathrm{dist}(Tw_k, \mathbb{R}w_k) \overset{k\to\infty}{\longrightarrow} 0.$$

Therefore there exists a sequence $(\lambda_k)_{k\in\mathbb{N}}$ in $\mathbb{R}$ such that

$$\|Tw_k - \lambda_k w_k\| \overset{k\to\infty}{\longrightarrow} 0.$$

The sequence $(\lambda_k)_{k\in\mathbb{N}}$ is bounded, since T is a bounded operator and $\|w_k\| > \delta$ for all k therefore it has a convergent subsequence. We may assume that $\lim_k \lambda_k = \lambda$ for some $\lambda \in \mathbb{R}$. We also have that $\|Tw_k\| = \|v_{2k-1} - v_{2k}\| \geq \|v_{2k}\| > \frac{\theta}{2}$ for each k and $(w_k)_{k\in\mathbb{N}}$ is a bounded sequence therefore $\lambda \neq 0$.

We set $s_k = \frac{w_k}{\|w_k\|}$, pass to a subsequence in order to have

$$\sum_{k=1}^{\infty} \|Ts_k - \lambda s_k\| < \frac{|\lambda|}{2}$$

and consider the subspace $F = \overline{\mathrm{span}}\{s_k : k \in \mathbb{N}\}$ of X_G. We claim that the restriction of T on F is an isomorphic embedding and hence the operator T is not strictly singular.

Indeed, let $s \in F$ with $\|s\| = 1$, $s = \sum_{k=1}^{\infty} b_k s_k$. Then $|b_k| \leq 1$ for each k thus

$$\|Ts - \lambda s\| \leq \sum_{k=1}^{\infty} |b_k| \|Ts_k - \lambda s_k\| \leq \frac{|\lambda|}{2}$$

hence $\|Ts\| \geq \frac{|\lambda|}{2}$. □

DEFINITION 9.5. A Banach space E is said to be somewhat reflexive if every infinite dimensional subspace of E contains a further infinite dimensional reflexive subspace.

PROPOSITION 9.6. *Suppose that the ground set G is $\mathcal{S}_2$ bounded and let $(z_n)_{n\in\mathbb{N}}$ be a normalized weakly null block sequence in $X_G = T[G, (\mathcal{S}_{n_j}, \frac{1}{m_j})_j, D]$. Then there exists $L \in [\mathbb{N}]$ such that the space $Z = \overline{\text{span}}\{z_n : n \in L\}$ is reflexive.*

Proof. We use the following notation. For a weakly null block sequence $\vec{w} = (w_n)_{n\in\mathbb{N}}$, $x^* \in \mathrm{B}_{X_G^*}$ and $\varepsilon > 0$ we denote by $Q(\vec{w}, x^*, \varepsilon)$ the subset of the natural numbers

$$Q(\vec{w}, x^*, \varepsilon) = \{\min \operatorname{supp} w_n : n \in \mathbb{N} \text{ such that } |x^*(w_n)| > \varepsilon\}.$$

Every weakly null block sequence in X_G is also weakly null in Y_G. Hence the definition of $\mathcal{S}_2$ boundness yields that if $(w_n)_{n\in\mathbb{N}}$ is a weakly null block sequence in X_G and $\varepsilon > 0$ there exists an $L \in [\mathbb{N}]$ such that $Q((w_n)_{n\in L}, g, \varepsilon) \in \mathcal{S}_2$ for every $g \in \overline{G}^{w^*}$ (see Remark 6.4, 3).

We construct a sequence $L_0 \supset L_1 \supset L_2 \supset L_3 \supset \cdots\cdots$ of infinite subsets of $\mathbb{N}$ such that

$$Q((z_n)_{n\in L_k}, g, \frac{1}{2^k}) \in \mathcal{S}_2$$

for each $k = 1, 2, \dots$ and every $g \in \overline{G}^{w^*}$.

The construction is inductive. Set $L_0 = \mathbb{N}$. Suppose that $L_0 \supset L_1 \supset L_2 \supset \cdots \supset L_{k-1}$ have been constructed such that the inductive assumption is fulfilled. We set

$$\mathcal{A}_k = \{L \in [L_{k-1}] : Q((z_n)_{n\in L}, g, \frac{1}{2^k}) \in \mathcal{S}_2 \text{ for every } g \in \overline{G}^{w^*}\} \text{ and}$$

$$\mathcal{B}_k = [L_{k-1}] \setminus \mathcal{A}_k.$$

It is clear that the set $\mathcal{B}_k$ is open in the pointwise topology of $[\mathbb{N}]$ thus by infinite Ramsey Theorem ([**40**], [**23**]) there exists an $L_k \in [L_{k-1}]$ such that either $[L_k] \subset \mathcal{A}_k$ or $[L_k] \subset \mathcal{B}_k$. Since the ground set G is $\mathcal{S}_2$ bounded it not possible to have $[L_k] \subset \mathcal{B}_k$; therefore $[L_k] \subset \mathcal{A}_k$. It is clear that the set L_k satisfies the desired property. The inductive construction is complete.

Let $L = \{l_1 < l_2 < l_3 < \cdots\}$ be such that $l_n \in L_n$ for $n = 1, 2, \dots$ and set $y_n = z_{l_n}$ for all n. We claim that the space $Z = \overline{\text{span}}\{z_n : n \in L\} = \overline{\text{span}}\{y_n : n \in \mathbb{N}\}$ is reflexive.

It is enough to show that the Schauder basis $(y_n)_{n\in\mathbb{N}}$ of Z is boundedly complete and shrinking. The first is trivial since $(y_n)_{n\in\mathbb{N}}$ is a block sequence of the boundedly complete Schauder basis $(e_n)_{n\in\mathbb{N}}$ of X_G. To see that $(y_n)_{n\in\mathbb{N}}$ is shrinking we show that

$$\|g|\overline{\text{span}}\{y_i : i \geq n\}\| \overset{n\to\infty}{\longrightarrow} 0$$

for every $g \in \overline{G}^{w^*}$. Accomplishing this, Proposition 6.10 yields that $\|x^*|\overline{\text{span}}\{y_i : i \geq n\}\| \overset{n\to\infty}{\longrightarrow} 0$ for every $x^* \in X_G^*$.

Let $g \in \overline{G}^{w^*}$. We consider a partition of the natural numbers in the following manner

$$F_1 = \{1\} \cup \{i \geq 1 : |g(y_i)| > \frac{1}{2}\}$$

$$F_k = \left(\{k\} \cup \{i \geq k : \frac{1}{2^{k-1}} \geq |g(y_i)| > \frac{1}{2^k}\right) \setminus \bigcup_{i=1}^{k-1} F_i$$

We observe that for each k the family $\{y_i : i \in F_k \setminus \{k\}\}$ is $\mathcal{S}_2$ admissible and also that if $i \in F_k$ then $|g(y_i)| \leq \frac{1}{2^{k-1}}$.

Let $l \in \mathbb{N}$. We show that for $n > \max(\bigcup_{i=1}^{l} F_i)$ we have $\|g|\overline{\operatorname{span}}\{y_i : i \geq n\}\| \leq \frac{10}{2^l}$. Consider $y \in \overline{\operatorname{span}}\{y_i : i \geq n\}$ with $\|y\| \leq 1$. Then $y = \sum_{i=n}^{\infty} \lambda_i y_i$ where $|\lambda_i| \leq 1$ for each i.

We show that for each k we have

$$\sum_{i \in F_k} |\lambda_i| \leq 5.$$

Indeed, for each i let $f_i \in B_{X_G^*}$ with $\operatorname{ran} f_i \subset \operatorname{ran} y_i$ such that $f_i(\lambda_i y_i) = \|\lambda_i y_i\|$. Then the family $\{f_i : i \in F_k \setminus \{k\}\}$ is $\mathcal{S}_2$ admissible. Since $m_2 = 4$ and the norming set D of the space X_G is 2 closed, the functional $f = \frac{1}{4} \sum_{i \in F_k \setminus \{k\}} f_i$ belongs to $B_{X_G^*}$.

It follows that

$$1 \geq \|y\| \geq f(y) = \frac{1}{4} \sum_{i \in F_k \setminus \{k\}} \|\lambda_i y_i\| = \frac{1}{4} \sum_{i \in F_k \setminus \{k\}} |\lambda_i|.$$

Therefore

$$\sum_{i \in F_k} |\lambda_i| \leq |\lambda_k| + \sum_{i \in F_k \setminus \{k\}} |\lambda_i| \leq 1 + 4 = 5.$$

We conclude that

$$|g(y)| \leq \sum_{i=n}^{\infty} |\lambda_i||g(y_i)| \leq \sum_{k=l+1}^{\infty} \sum_{i \in F_k} |\lambda_i||g(y_i)| \leq \sum_{k=l+1}^{\infty} 5 \cdot \frac{1}{2^{k-1}} = \frac{10}{2^l}.$$

Therefore $\|g|\overline{\operatorname{span}}\{y_i : i \geq n\}\| \leq \frac{10}{2^l}$ for $n > \max(\bigcup_{i=1}^{l} F_i)$. The proof is complete. □

Proposition 9.6 immediately yields the following theorem.

THEOREM 9.7. *If the ground set G is $\mathcal{S}_2$ bounded then the space $X_G = T[G, (\mathcal{S}_{n_j}, \frac{1}{m_j})_j, D]$ is somewhat reflexive.*

The next theorem summarizes what has been proved up to this point.

THEOREM 9.8. *Let G be an $\mathcal{S}_2$ bounded subset of $c_{00}(\mathbb{N})$. Then for some double sequence $(n_j, m_j)_{j \in \mathbb{N}}$ and norming set D the space $X_G = T[G, (\mathcal{S}_{n_j}, \frac{1}{m_j})_j, D]$ has the following properties*

1. *The space X_G is asymptotic ℓ^1 and $(e_n)_{n \in \mathbb{N}}$ is a bimonotone boundedly complete Schauder basis.*
2. *The space X_G is Hereditarily Indecomposable and somewhat reflexive.*
3. *The space $(X_G)_* = \overline{\operatorname{span}}\{e_n^* : n \in \mathbb{N}\}$ is Hereditarily Indecomposable.*
4. *The space X_G^* is the norm closed linear span of the closure of G in the topology of pointwise convergence i.e. $X_G^* = \overline{\operatorname{span}}(\overline{G}^p)$.*

(5) *The space of bounded linear operators of* X_G *has the form*

$$\mathcal{L}(X_G) = \{\lambda I + S : \ \lambda \in \mathbb{R}, \ \ S \textit{ is strictly singular and weakly compact operator } \}.$$

Proof. (1) has been proved in Proposition 3.11. Theorems 7.3, 9.7 yield (2), while (3) follows from Theorem 8.3. Also (4) is proved in Lemma 6.10 and, finally, (5) has been shown in Theorems 9.2 and 9.4. □

CHAPTER 10

Defining G to obtain a nonseparable H.I. space X_G^*

In this chapter we define an $\mathcal{S}_2$ bounded set G_{NS} which produces a space $X_{G_{NS}}$ such that $X_{G_{NS}}^*$ is a nonseparable H.I. Banach space. The set G_{NS} will be defined over the binary tree. We also show that $X_{G_{NS}}^*$ is somewhat reflexive and we study the structure of $\mathcal{L}(X_{G_{NS}}^*)$. As result we obtain that every operator $T: X_{G_{NS}}^* \longrightarrow X_{G_{NS}}^*$ is of the form $\lambda I + R$ with R weakly compact and hence R has separable range.

The *binary tree* $\mathcal{D}$ is the uniquely rooted tree of height ω each node of which has exactly two immediate successors. We consider the standard representation of the binary tree as the set of all finite sequences of 0's and 1's partially ordered with the relation

$$a \preceq b \iff a \text{ is an initial segment of } b.$$

For every $a \in \mathcal{D}$ the *rank* $|a|$ of a is defined by $|a| = \#\{x \in \mathcal{D} : x \preceq a, \quad x \neq a\}$ and coincides with the length of the sequence a.

Segments of $\mathcal{D}$ are the totally ordered subsets of $\mathcal{D}$ while *branches* of $\mathcal{D}$ are its maximal segments. The set of all branches of $\mathcal{D}$ is denoted by $\mathcal{B}$ and has the cardinality of the continuum.

We also consider the *lexicographic order* of $\mathcal{D}$ defined through the bijection $h: \mathcal{D} \longrightarrow \mathbb{N}$ described by the following rule:

$$h(a) = \begin{cases} 1 & \text{if } |a| = 0 \\ 2^{|a|} + \sum\limits_{j=1}^{|a|} a(j)2^{|a|-j} & \text{if } |a| \geq 1. \end{cases}$$

The function h endows the set of natural numbers with a partial order $\preceq$ such that $(\mathbb{N}, \preceq)$ coincides with the binary tree and the natural order of $\mathbb{N}$ is the lexicographic order of the binary tree. Clearly the function h induces a homomorphism between $c_{00}(\mathbb{N})$ and $c_{00}(\mathcal{D})$.

In the sequel by the term segments we shall mean the segments of the binary tree while by the term intervals we mean the intervals of $\mathbb{N}$ in the natural order. For a segment s we denote by s^* the characteristic function χ_s which belongs in general to $\ell^\infty(\mathbb{N})$. If s is a finite segment then s^* belongs to $c_{00}(\mathbb{N})$.

REMARK 10.1. A direct consequence of the definition of the partial order $\preceq$ on $\mathbb{N}$ is that for any segment s and any interval E the intersection $E \cap s$ is also a segment.

DEFINITION 10.2. We define the set G_{NS} as

$$\begin{aligned} G_{NS} &= \Big\{ \sum_{i=1}^{d} \varepsilon_i s_i^* : \varepsilon_i \in \{-1,1\}, \ i = 1,2,\ldots,d, \ (s_i)_{i=1}^d \text{ are pairwise disjoint} \\ &\qquad \text{finite segments with } \{\min s_1, \min s_2, \ldots, \min s_d\} \in \mathcal{S}, \ d \in \mathbb{N} \Big\}. \end{aligned}$$

where $\mathcal{S} = \{F \subset \mathbb{N} : \#F \leq \min F\} \cup \{\emptyset\}$ is the first Schreier family.

LEMMA 10.3. (1) The set G_{NS} is ground.
(2) The closure of the set G_{NS} in the topology of pointwise convergence is

$$\overline{G_{NS}}^p = \Big\{\sum_{i=1}^{d} \varepsilon_i s_i^* : \varepsilon_i \in \{-1,1\},\ i=1,2,\ldots,d, (s_i)_{i=1}^d \text{ are pairwise disjoint segments with } \{\min s_1, \min s_2, \ldots, \min s_d\} \in \mathcal{S},\ d \in \mathbb{N}\Big\}.$$

Proof. It follows readily (see also Remark 10.1) that the set G_{NS} is ground (Definition 3.1). To prove the equality (2), let $g \in \overline{G_{NS}}^p$ and we show that it is of the form described in (2) (the other inclusion is trivial). Let $(g_k)_{k\in\mathbb{N}}$ be a sequence in G_{NS} such that $g_k \xrightarrow{p} g$. Set $d = \min \operatorname{supp} g$. Since $g_k(n) \xrightarrow{k\to\infty} 0$ for $n = 1, \ldots, d-1$ and $g_k(d) \xrightarrow{k\to\infty} g(d) \neq 0$ and from the fact that each $g_k(n)$ belongs to $\{-1, 0, 1\}$ we may assume, passing to a subsequence, that $\min \operatorname{supp} g_k = d$ for all $k \in \mathbb{N}$. Let $g_k = \sum_{i=1}^{d} \varepsilon_i^k (s_i^k)^*$ for $k = 1, 2\ldots$

Observe that the limit of a sequence of finite segments in the pointwise topology is a segment which can be either finite or infinite. We thus may select $L_1 \in [\mathbb{N}]$, $\varepsilon_1 \in \{-1, 1\}$ and a segment s_1 such that $\varepsilon_1 s_1^*$ is the pointwise limit of the sequence $(\varepsilon_1^k (s_1^k)^*)_{k\in L_1}$. After d consecutive applications of the same argument we may select infinite sets of natural numbers $L_1 \supset L_2 \supset \cdots \supset L_d$, $\varepsilon_1, \varepsilon_2, \ldots \varepsilon_d \in \{-1, 1\}$ and disjoint segments $s_1, s_2, \ldots, s_d$ such that $(\varepsilon_i^k (s_i^k)^*)_{k\in L_i} \xrightarrow{p} \varepsilon_i s_i^*$ for $i = 1, 2, \ldots, d$. We deduce that $g = \sum_{i=1}^{d} \varepsilon_i s_i^*$. □

We next show that G_{NS} is $\mathcal{S}_2$ bounded. We remind at this point that $Y_{G_{NS}}$ denotes the completion of the space $(c_{00}, \|\ \|_{G_{NS}})$.

PROPOSITION 10.4. The space $Y_{G_{NS}}$ does not contain isomorphically ℓ^1.

Proof. It is enough to show that every bounded sequence in $Y_{G_{NS}}$ has a weakly Cauchy subsequence. Let $(x_n)_{n\in\mathbb{N}}$ be a bounded sequence in $Y_{G_{NS}}$. By Rainwater's Theorem is is enough to find a subsequence $(x_n)_{n\in M}$ such that $(f(x_n))_{n\in M}$ is convergent for every $f \in \operatorname{Ext}(B_{Y^*_{G_{NS}}})$. Since $B_{Y^*_{G_{NS}}} = \overline{\operatorname{conv}(G_{NS})}^{w^*}$ by Milman's Theorem we get that $\operatorname{Ext}(B_{Y^*_{G_{NS}}}) \subset \overline{G_{NS}}^{w^*}$. Therefore, by the form of elements of $\overline{G_{NS}}^{w^*}$ it is enough to show that there exists $M \in [\mathbb{N}]$ such that $(s^*(x_n))_{n\in M}$ is convergent for every segment s. This is done in the following two lemmas. □

LEMMA 10.5. Let $(x_n)_{n\in\mathbb{N}}$ be a bounded sequence in $Y_{G_{NS}}$ and $\varepsilon > 0$. Then there exists a finite set $\{s_1, s_2, \ldots, s_k\}$ of pairwise disjoint segments and an $L \in [\mathbb{N}]$ such that

$$\limsup_{n\in L} |s^*(x_n)| \leq \varepsilon$$

for every segment s with $s \cap (\bigcup_{i=1}^{k} s_i) = \emptyset$.

Proof. Assume the contrary. Then for every finite set $\{t_1, t_2, \ldots, t_m\}$ of pairwise disjoint segments and every $L \in [\mathbb{N}]$ there exists a segment t with $t \cap (\bigcup_{i=1}^{m} t_i) = \emptyset$ such that $\limsup_{n \in L} |t^*(x_n)| > \varepsilon$.

Using this fact we may inductively construct a sequence $(s_j)_{j \in \mathbb{N}}$ of pairwise disjoint segments of $\mathcal{D}$ and a decreasing sequence $(L_j)_{j \in \mathbb{N}}$ of infinite subsets of the natural numbers such that

$$|s_j^*(x_n)| > \varepsilon \qquad \forall j \in \mathbb{N} \quad \forall n \in L_j.$$

We set $r = \sup\{\|x_n\| : n \in \mathbb{N}\}$ and choose $k \in \mathbb{N}$ with $k > \frac{r}{\varepsilon}$. Since the segments $s_1, s_2, s_3, \ldots$ are pairwise disjoint we may choose an $i_0 \in \mathbb{N}$ such that $\min s_i \geq k$ for every $i > i_0$.

Let $n \in L_{i_0+k}$. Then $n \in L_{i_0+t}$ for each $t \in \{1, 2, \ldots, k\}$ i.e. $|s_{i_0+t}^*(x_n)| > \varepsilon$ for each $t \in \{1, 2, \ldots, k\}$. Setting $\varepsilon_t = \operatorname{sgn}(s_{i_0+t}^*(x_n))$ we have that

$$f = \sum_{t=1}^{k} \varepsilon_t s_{i_0+t}^* \in \overline{G_{NS}}^{w^*}$$

thus $f \in B_{Y_{G_{NS}}^*}$. It follows that

$$r \geq \|x_n\| \geq f(x_n) = \sum_{t=1}^{k} \varepsilon_t s_{i_0+t}^*(x_n) = \sum_{t=1}^{k} |s_{i_o+t}^*(x_n)| > k\varepsilon$$

a contradiction of the choice of k. □

LEMMA 10.6. *Let $(x_n)_{n \in \mathbb{N}}$ be a bounded sequence in $Y_{G_{NS}}$. There exists a subsequence $(x_n)_{n \in M}$ of $(x_n)_{n \in \mathbb{N}}$ such that for every segment s the sequence $(s^*(x_n))_{n \in M}$ is convergent.*

Proof. From Lemma 10.5 we may inductively construct a decreasing sequence $(L_k)_{k \in \mathbb{N}}$ of infinite subsets of the natural numbers and a sequence $(F_k)_{k \in \mathbb{N}}$ of finite sets of pairwise disjoint segments, $F_k = \{s_1^k, s_2^k, \ldots, s_{m_k}^k\}$ such that for each $k \in \mathbb{N}$ and segment s with $s \cap (\bigcup_{i=1}^{m_k} s_i^k) = \emptyset$ we have that

$$\limsup_{n \in L_k} |s^*(x_n)| < \frac{1}{k}.$$

Let F be the countable set consisting of all the finite segments of $\mathcal{D}$ and all subsegments of segments contained in $\bigcup_{k=1}^{\infty} F_k$. We choose a diagonal set L of the decreasing sequence $(L_k)_{k \in \mathbb{N}}$. Since $\{s^* : s \in F\}$ is a countable subset of $B_{Y_{G_{NS}}^*}$ and the sequence $(x_n)_{n \in L}$ is bounded we may choose, by a diagonal argument, an $M \in [L]$ such that the sequence $(s^*(x_n))_{n \in M}$ is convergent for every $s \in F$. It remains to show that the sequence $(s^*(x_n))_{n \in M}$ is convergent for every segment s.

Let s be a segment. We show that $(s^*(x_n))_{n \in M}$ is a Cauchy sequence. Let $\varepsilon > 0$ and choose $k \in \mathbb{N}$ with $\frac{1}{k} < \frac{\varepsilon}{4}$. We have

$$s = \bigcup_{i=1}^{m_k} (s_i^k \cap s) \cup (s \setminus \bigcup_{i=1}^{m_k} s_i^k)$$

and $s \setminus \bigcup_{i=1}^{m_k} s_i^k$ can be written as the finite union of pairwise disjoint segments with at most one of them being infinite. We assume that one of them is infinite and let

$$s \setminus \bigcup_{i=1}^{m_k} s_i^k = (\bigcup_{j=1}^{n_k} t_j) \cup t$$

where $t_1, t_2, \ldots, t_{n_k}, t$ are pairwise disjoint segments, t is infinite while $t_1, t_2, \ldots, t_{n_k}$ are finite.

It is clear that $t \cap (\bigcup_{i=1}^{m_k} s_i^k) = \emptyset$ thus $\limsup_{n \in L_k} |t^*(x_n)| < \frac{1}{k}$ and by the construction of M as a (subset of a) diagonal set we have that $\limsup_{n \in M} |t^*(x_n)| < \frac{1}{k}$ thus we may choose an $m_a \in M$ such that

$$|t^*(x_n)| < \frac{1}{k} \quad \forall n \in M \text{ with } n \geq m_a.$$

Since $s_i^k \cap s \in F$ for each $i = 1, 2, \ldots, m_k$ and $t_j \in F$ for each $j = 1, 2, \ldots, n_k$ the sequences $\big((s_i^k \cap s)^*(x_n)\big)_{n \in M}$ and $\big(t_j^*(x_n)\big)_{n \in M}$ are convergent. Thus we may select an $m_b \in M$ such that

$$\sum_{i=1}^{m_k} |(s_i^k \cap s)^*(x_m) - (s_i^k \cap s)^*(x_n)| + \sum_{j=1}^{n_k} |t_j^*(x_m) - t_j^*(x_n)| < \frac{\varepsilon}{2}$$

for each $m, n \in M$ with $m > n \geq m_b$.

Therefore for each $m, n \in M$ with $m > n \geq \max\{m_a, m_b\}$ we have that

$$|s^*(x_m) - s^*(x_n)| < \frac{\varepsilon}{2} + \frac{2}{k} < \varepsilon.$$

Thus $(s^*(x_n))_{n \in M}$ is a Cauchy, and hence convergent, sequence. □

LEMMA 10.7. Let $\varepsilon > 0$ and $y \in Y_{G_{NS}}$ such that $\min \operatorname{supp} y \geq [\frac{\|y\|}{\varepsilon}] + 1$. If $s_1, s_2, \ldots, s_t$ are pairwise disjoint segments such that $s_i \subset \operatorname{ran} y$ and $|s_i^*(y)| > \varepsilon$ for each $i = 1, 2, \ldots, t$ then $t \leq \frac{\|y\|}{\varepsilon}$.

Proof. Suppose that $t > \frac{\|y\|}{\varepsilon}$ and choose $A \subset \{1, 2, \ldots, t\}$ with $\#A = [\frac{\|y\|}{\varepsilon}] + 1$. Set $\varepsilon_j = \operatorname{sgn}(s_j^*(y))$ for each $j \in A$ and $f = \sum_{j \in A} \varepsilon_j s_j^*$. We have $f \in \overline{G_{NS}}^p \subset B_{Y_{G_{NS}}^*}$. Indeed, the segments $(s_j)_{j \in A}$ are pairwise disjoint while we have that $\min s_j \geq \min \operatorname{supp} y \geq [\frac{\|y\|}{\varepsilon}] + 1 = \#A$. Thus

$$\|y\| \geq f(y) = \sum_{j \in A} \varepsilon_j s_j^*(y) = \sum_{j \in A} |s_j^*(y)| > \varepsilon(\#A) = \varepsilon([\frac{\|y\|}{\varepsilon}] + 1) > \|y\|.$$

a contradiction.

Therefore $t \leq \frac{\|y\|}{\varepsilon}$. □

LEMMA 10.8. Let $(y_n)_{n \in \mathbb{N}}$ be a weakly null block sequence in $Y_{G_{NS}}$. Then for every $\varepsilon > 0$ there exists an $L \in [\mathbb{N}]$ such that for every segment t we have that

$$\{n \in L : \ |t^*(y_n)| > \varepsilon\} \leq 2$$

Proof. We set

$$\mathcal{A} = \Big\{\{k_1,k_2,k_3\}\in[\mathbb{N}]^3 : \text{ there exists a segment } s \text{ such that } |s^*(y_{k_l})|>\varepsilon \text{ for } l=1,2,3\Big\}$$

and

$$\mathcal{B}=[\mathbb{N}]^3\setminus\mathcal{A}.$$

It follows from Ramsey's Theorem that there exists an $L\in[\mathbb{N}]$ such that $[L]^3\subset\mathcal{A}$ or $[L]^3\subset\mathcal{B}$. The conclusion of the lemma is exactly that there exists an $L\in[\mathbb{N}]$ such that $[L]^3\subset\mathcal{B}$. Therefore it is enough to exclude the possibility to exist an $L\in[\mathbb{N}]$ with $[L]^3\subset\mathcal{A}$.

Suppose that there exists an $L\in[\mathbb{N}]$, $L=\{l_1<l_2<l_3<\cdots\}$ such that $[L]^3\subset\mathcal{A}$. We set $r=\sup_n\|y_n\|$. We may assume that $\min L>\frac{2r}{\varepsilon}$.

For each $k\geq 3$ we apply the following procedure. We consider the set

$$T_k = \Big\{\operatorname{ran}(y_{l_2}+y_{l_{k-1}})\cap s:\ s \text{ is a segment , such that } |s^*(y_{l_1})|>\varepsilon,\ |s^*(y_{l_k})|>\varepsilon, \text{ and there exists a } t\in\{2,\dots,k-1\} \text{ such that } |s^*(y_{l_t})|>\varepsilon\Big\}.$$

We claim that $\#T_k\leq\frac{r}{\varepsilon}$. Indeed, let $t_1,t_2,\dots,t_m$ be pairwise different elements of T_k and for each $i=1,2,\dots,m$ let s_i be a segment such that $\operatorname{ran}(y_{l_2}+y_{l_{k-1}})\cap s_i = t_i$ and $|s_i(y_{l_k})|>\varepsilon$. The segments $r_i=(\operatorname{ran}y_{l_k})\cap s_i$ $i=1,2,\dots,m$ are pairwise disjoint. We also have that $|r_i^*(y_{l_k})|>\varepsilon$ and $r_i\subset\operatorname{ran}y_{l_k}$ for each $i=1,2,\dots,m$. From the fact that $\|y_{l_k}\|\geq|r_i^*(y_{l_k})|>\varepsilon$ we get that

$$\min\operatorname{supp}y_{l_k}\geq l_k>l_1>\frac{2r}{\varepsilon}\geq\frac{2\|y_{l_k}\|}{\varepsilon}>[\frac{\|y_{l_k}\|}{\varepsilon}]+1.$$

Lemma 10.7 yields that $m\leq\frac{\|y_{l_k}\|}{\varepsilon}$ thus $m\leq\frac{r}{\varepsilon}$. We conclude that $\#T_k\leq\frac{r}{\varepsilon}$. We set $m_0=[\frac{r}{\varepsilon}]$. Then

$$T_k=\{s_1^k,s_2^k,\dots,s_{t_k}^k\}$$

for some $t_k\leq m_0$.

For $k\geq 3$ and $i=1,2,\dots,t_k$ we set

$$A_i^k=\Big\{j\in\{2,\dots,k-1\}:\ |(s_i^k)^*(y_{l_j})|>\varepsilon\Big\}$$

while for $t_k<i\leq m_0$ we set $A_i^k=\emptyset$. Hence

$$\{2,\dots,k-1\}=\bigcup_{i=1}^{m_0}A_i^k. \tag{34}$$

We consider the following partition of $[\mathbb{N}\setminus\{1\}]^2$. For $j=1,2,\dots,m_0$ we set

$$B_j=\Big\{\{p,q\}\in[\mathbb{N}\setminus\{1\}]^2:\ p<q \text{ and } p\in A_j^q\Big\}.$$

Observe that (34) yields that $[\mathbb{N}\setminus\{1\}]^2=\bigcup_{j=1}^{m_0}B_j$ thus by Ramsey's theorem there exist an $M\in[\mathbb{N}\setminus\{1\}]$ and a $j_0\in\{1,2,\dots,m_0\}$ such that $[M]^2\subset B_{j_0}$.

Let $M = \{m_1 < m_2 < m_3 < \cdots\}$. Then for each $t \in \mathbb{N}$ and $i \leq t$ we have that $\{m_i, m_{t+1}\} \in B_{j_0}$ thus $m_i \in A_{j_0}^{m_{t+1}}$. So there exists a segment s_t such that

$$|s_t^*(y_{m_i})| > \varepsilon \qquad \forall i = 1, 2, \ldots, t.$$

The sequence $(s_t^*)_{t\in\mathbb{N}}$ has a w^* convergent subsequence; its w^* limit is of the form s^* for some infinite segment s. We deduce that $|s^*(y_{m_i})| > \varepsilon$ for all i which contradicts to the assumption that the sequence $(y_n)_{n\in\mathbb{N}}$ is weakly null. □

LEMMA 10.9. *Let $(x_n)_{n\in\mathbb{N}}$ be a weakly null block sequence in $Y_{G_{NS}}$. Then for every $\varepsilon > 0$ there exists an $L \in [\mathbb{N}]$ such that the sequence $(x_n)_{n\in L}$ is (G_{NS}, ε) Schreier.*
(i.e. for every $g \in G_{NS}$ the set $\{t_k : \ k \in \mathbb{N}$ such that $|g(x_k)| > \varepsilon\}$ belongs to the second Schreier family $\mathcal{S}_2$ (where $t_k = \min \operatorname{supp} x_k$ for each k).)

Proof. We construct, with repeated use of Lemma 10.8, a decreasing sequence $(L_k)_{k\in\mathbb{N}}$ of infinite subsets of the natural numbers such that the following conditions hold

(i) $\min \operatorname{supp} x_{l_1} \geq 3$.
(ii) $l_k = \min L_k \notin L_{k+1}$ for each $k \in \mathbb{N}$.
(iii) For each $k \in \mathbb{N}$ if $p_k = \max \operatorname{supp} x_{l_k}$ then for every segment s we have that
$$\#\{n \in L_{k+1} : \ |s^*(x_n)| > \frac{\varepsilon}{p_k}\} \leq 2.$$

We set $L = \{l_1, l_2, l_3, \ldots\}$ and we claim that the sequence $(x_n)_{n\in L}$ is (G_{NS}, ε) Schreier.

Indeed, let $g = \sum_{i=1}^{d} \varepsilon_i s_i^* \in G_{NS}$ where the segments $s_1, s_2, \ldots, s_d$ of $\mathcal{D}$ are nonempty and pairwise disjoint, $d \leq \min s_i$ and $\varepsilon_i \in \{-1, 1\}$ for each $i = 1, 2, \ldots, d$. Set

$$l_{i_0} = \min\{n \in L : \ \operatorname{supp} x_n \cap \operatorname{supp} g \neq \emptyset\}.$$

Observe that $d \leq p_{i_0}$. We set

$$F = \{n \in L : \ |g(x_n)| > \varepsilon\}.$$

We have that $F \subset \{l_{i_0}, l_{i_0+1}, l_{i_0+2}, \ldots\}$ thus $F \setminus \{l_{i_0}\} \subset L_{i_0+1}$. Also (iii) yields that for each $i = 1, 2, \ldots, d$ the set $F_i = \{n \in L_{i_0+1} : \ |s_i^*(x_n)| > \frac{\varepsilon}{p_{i_0}}\}$ has at most two elements.

We observe that

$$F \setminus \{l_{i_0}\} \subset \bigcup_{i=1}^{d} F_i.$$

Indeed if $n \in L_{i_0+1}$ and $n \notin \bigcup_{i=1}^{d} F_i$ then by the inductive construction $|s_i^*(x_n)| \leq \frac{\varepsilon}{p_{i_0}}$ for each $i = 1, 2, \ldots, d$ thus $|\sum_{i=1}^{d} \varepsilon_i s_i^*(x_n)| \leq d\frac{\varepsilon}{p_{i_0}}$ and it follows that $|g(x_n)| \leq \varepsilon$ therefore $n \notin F$.

We conclude that $\#(F \setminus \{l_{i_0}\}) \leq 2d$. Also $t_n = \min \operatorname{supp} x_n > p_{i_0} \geq d$ for each $n \in F \setminus \{l_{i_0}\}$ hence the set $\{t_n : \ n \in F \setminus \{l_{i_0}\}\}$ is the union of two sets belonging to the first Schreier family $\mathcal{S}_1$. Since $t_{l_{i_0}} \geq 3$ the set $\{t_n : \ n \in F\}$ is the union of three sets of $\mathcal{S}_1$ and its minimum is greater or equal to 3. It follows that

$$\{t_n : \ n \in F\} \in \mathcal{S}_2.$$

Therefore the sequence $(x_n)_{n\in L}$ is (G_{NS},ε) Schreier. □

Proposition 10.4 and Lemma 10.9 immediately yield the following:

PROPOSITION 10.10. The set G_{NS} is $\mathcal{S}_2$ bounded.

In the rest of this chapter $X_{G_{NS}}$ stands for the space

$$X_{G_{NS}} = T\left[G_{NS}, \left(\mathcal{S}_{n_j}, \frac{1}{m_j}\right)_j, D\right].$$

PROPOSITION 10.11. Let $\mathcal{B}$ denote the set of all branches of the binary tree $\mathcal{D}$. Then we have the following:

(i) $X^*_{G_{NS}} = \overline{\operatorname{span}}(\{e^*_n : n \in \mathbb{N}\} \cup \{b^* : b \in \mathcal{B}\})$.
(ii) $X^*_{G_{NS}}/(X_{G_{NS}})_* = \overline{\operatorname{span}}\{b^* + (X_{G_{NS}})_* : b \in \mathcal{B}\} = c_0(\mathcal{B})$.
(iii) The space $X^{**}_{G_{NS}}$ is isomorphic to $X_{G_{NS}} \oplus \ell^1(\mathcal{B})$.

Proof. (i) By Lemma 10.3(ii) we have that the set $\overline{G_{NS}}^{w^*}$ is equal to the set

$$\overline{G_{NS}}^{p} = \Big\{\sum_{i=1}^{d} \varepsilon_i s_i^* : \varepsilon_i \in \{-1,1\},\ i = 1,2,\dots,d,\ (s_i)_{i=1}^d \text{ are pairwise disjoint segments and } \{\min s_1, \min s_2, \dots, \min s_d\} \in \mathcal{S}_1,\ d \in \mathbb{N}\Big\}.$$

Therefore, by Proposition 6.10 we get that $X^*_{G_{NS}} = \overline{\operatorname{span}}(\{e^*_n : n \in \mathbb{N}\} \cup \{b^* : b \in \mathcal{B}\})$.

(ii) By Proposition 3.11 the basis $(e_n)_{n\in\mathbb{N}}$ is boundedly complete and the space $(X_{G_{NS}})_* = \overline{\operatorname{span}}\{e^*_n : n \in \mathbb{N}\}$ is the predual of $X_{G_{NS}}$. It follows from (i) that $X^*_{G_{NS}}/(X_{G_{NS}})_* = \overline{\operatorname{span}}\{b^* + (X_{G_{NS}})_* : b \in \mathcal{B}\}$ so it remains to show that $\overline{\operatorname{span}}\{b^* + (X_{G_{NS}})_* : b \in \mathcal{B}\} = c_0(\mathcal{B})$. It is enough to show that if $b_1, b_2, \dots, b_n$ are pairwise different elements of $\mathcal{B}$ and $\varepsilon_1, \varepsilon_2, \dots, \varepsilon_n \in \{-1,1\}$ then

$$\|\varepsilon_1(b_1^* + (X_{G_{NS}})_*) + \varepsilon_2(b_2^* + (X_{G_{NS}})_*) + \cdots + \varepsilon_n(b_n^* + (X_{G_{NS}})_*)\| = 1.$$

For $k = 1,2,\dots$ we denote by E_k the interval $E_k = \{n \in \mathbb{N} : n \geq 2^k\}$. Since $b_1, b_2, \dots, b_n$ are pairwise different we may select a $k_0 \geq n$ such that $b_1 \cap E_k, b_2 \cap E_k, \dots, b_n \cap E_k$ are incomparable segments for all $k \geq k_0$. Thus

$$\begin{aligned}
&\|\varepsilon_1(b_1^* + (X_{G_{NS}})_*) + \cdots + \varepsilon_n(b_n^* + (X_{G_{NS}})_*)\| \\
&= \operatorname{dist}(\varepsilon_1 b_1^* + \cdots + \varepsilon_n b_n^*,\ (X_{G_{NS}})_*) \\
&= \operatorname{dist}(\varepsilon_1 b_1^* + \cdots + \varepsilon_n b_n^*,\ \overline{\operatorname{span}}\{e_j^* : j \in \mathbb{N}\}) \\
&= \operatorname{dist}(\varepsilon_1 b_1^* + \cdots + \varepsilon_n b_n^*,\ \operatorname{span}\{e_j^* : j \in \mathbb{N}\}) \\
&= \lim_k \operatorname{dist}(\varepsilon_1 b_1^* + \cdots + \varepsilon_n b_n^*,\ \operatorname{span}\{e_j^* : j \leq 2^k - 1\}) \\
&= \lim_k \|\varepsilon_1(b_1 \cap E_k)^* + \cdots + \varepsilon_n(b_n \cap E_k)^*\| \\
&= 1.
\end{aligned}$$

The last equality holds since for every $k \geq k_0$ the functional $g_k = \sum_{i=1}^{n} \varepsilon_i (b_i \cap E_k)^*$ belongs to G_{NS}, thus $\|g_k\| \leq 1$, while for $n \in b_1 \cap E_k$ we have that $|g_k(e_n)| = 1$.

(iii) As is well known, for every Banach space Z the space Z^{***} is isomorphic to $Z^* \oplus (Z^{**}/Z)^*$. For $Z = (X_{G_{NS}})_*$ we get that the space $X^{**}_{G_{NS}}$ is isomorphic to the

space
$X_{G_{NS}} \oplus (X^*_{G_{NS}}/(X_{G_{NS}})_*)^* = X_{G_{NS}} \oplus (c_0(\mathcal{B}))^* = X_{G_{NS}} \oplus \ell^1(\mathcal{B}).$ □

THEOREM 10.12. *The space $X^*_{G_{NS}}$ is a nonseparable H.I. Banach space.*

Proof. The space $X_{G_{NS}}$, being H.I., contains no isomorphic copy of ℓ^1, the space $(X_{G_{NS}})_* = \overline{\text{span}}\{e^*_n : n \in \mathbb{N}\}$ is H.I. and the quotient space $X^*_{G_{NS}}/(X_{G_{NS}})_*$ is isometric to $c_0(\mathcal{B})$. From Corollary 1.5 we deduce that $X^*_{G_{NS}}$ is H.I. while it is clear that $X^*_{G_{NS}}$ is nonseparable. □

LEMMA 10.13. Let Z be a Banach space with a normalized Schauder basis $(e_i)_{i\in\mathbb{N}}$. If every normalized block sequence $(x_n)_{n\in\mathbb{N}}$, $x_n = \sum\limits_{i\in F_n} a_i e_i$ has the property

$$\liminf_n \|x_n\|_\infty > 0$$

(where $\|\sum a_i e_i\|_\infty = \sup|a_i|$) then the basis $(e_i)_{i\in\mathbb{N}}$ is equivalent to the standard basis of c_0.

Proof. Assume the contrary. Then we may construct a block sequence $(z_n)_{n\in\mathbb{N}}$, $z_n = \sum\limits_{i\in F_n} a_i e_i$, such that $\max\limits_{i\in F_n}|a_i| \le 1$ and $\|z_n\| \ge n$. Setting $x_n = \dfrac{z_n}{\|z_n\|}$ for each n, $(x_n)_{n\in\mathbb{N}}$ is a normalized block sequence and $\|x_n\|_\infty < \dfrac{1}{n}$ for all n, a contradiction. □

PROPOSITION 10.14. The space $X^*_{G_{NS}}$ is somewhat reflexive.

Proof. It is an easy observation that if X is a H.I. Banach space and Y is a subspace of X which is somewhat reflexive then the space X is also somewhat reflexive. Hence, in order to show that $X^*_{G_{NS}}$ is somewhat reflexive, it suffices to show that $(X_{G_{NS}})_*$ is somewhat reflexive.

Let also observe that if $f \in X^*_{G_{NS}} \setminus (X_{G_{NS}})_*$, $f = w^* - \sum f(e_n)e^*_n$ then $\limsup|f(e_n)| > 0$. Indeed $f + (X_{G_{NS}})_* = [f] \neq 0$ hence, by Proposition 10.11(ii), $[f] = \sum \lambda_b[b^*] \in c_0(\mathcal{B})$ with $\lambda_b \neq 0$ for some $b \in \mathcal{B}$. For such a b we have that $\lim\limits_{n\in b} f(e_n) = \lambda_b \neq 0$ therefore $\limsup|f(e_n)| > 0$.

Claim. If $(f_n)_{n\in\mathbb{N}}$ is a normalized block sequence in $(X_{G_{NS}})_*$ such that $\|f_n\|_\infty \longrightarrow 0$ then the space $F = \overline{\text{span}}\{f_n : n \in \mathbb{N}\}$ is reflexive.

Proof of the Claim. Suppose that the space F is not reflexive. Then there exists an f in the closure of B_F in the w^* topology of $X^*_{G_{NS}}$ which does not belong to $(X_{G_{NS}})_*$. Then f is the w^* limit of the series $\sum f(e_n)e^*_n$ with $\limsup|f(e_n)| > 0$. On the other hand f is equal to the w^* limit of some series $\sum \beta_n f_n$ with $|\beta_n| \le 1$ for all n. Since $\|f_n\|_\infty \longrightarrow 0$ we arrive to a contradiction.

To complete the proof it is enough to construct in every block subspace of $(X_{G_{NS}})_*$ a normalized block sequence $(f_n)_{n\in\mathbb{N}}$ with $\|f_n\|_\infty \longrightarrow 0$. This follows readily from Lemma 10.13 since $(X_{G_{NS}})_*$, being a H.I. space, contains no isomorphic copy of c_0. □

THEOREM 10.15. *There exists a Banach space $X_{G_{NS}}$ with a Schauder basis $(e_n)_{n\in\mathbb{N}}$ such that*

(1) *The space $X_{G_{NS}}$ is an asymptotic ℓ^1 space, and the Schauder basis $(e_n)_{n\in\mathbb{N}}$ is boundedly complete.*
(2) *The space $X_{G_{NS}}$ is Hereditarily Indecomposable and somewhat reflexive.*

(3) *The space* $(X_{G_{NS}})_* = \overline{\text{span}}\{e_n^* : n \in \mathbb{N}\}$ *is Hereditarily Indecomposable and somewhat reflexive.*

(4) *The quotient space* $X_{G_{NS}}^*/(X_{G_{NS}})_*$ *isometric to* $c_0(\Gamma)$ *with* $|\Gamma| = \mathfrak{c}$, *hence* $X_{G_{NS}}^*$ *is nonseparable.*

(5) *The space* $X_{G_{NS}}^*$ *is Hereditarily Indecomposable and somewhat reflexive.*

(6) *The space* $X_{G_{NS}}^{**}$ *is isomorphic to* $X_{G_{NS}} \oplus \ell^1(\Gamma)$.

(7) *The space* $\mathcal{L}(X_{G_{NS}})$ *of bounded linear operators on* $X_{G_{NS}}$ *has the form*

$$\mathcal{L}(X_{G_{NS}}) = \{\lambda I + S : \ \lambda \in \mathbb{R},\ S \text{ is strictly singular and weakly compact operator}\}.$$

(8) *Every bounded linear operator* $T : X_{G_{NS}}^* \longrightarrow X_{G_{NS}}^*$ *is of the form* $T = S^* + K$ *where* $S \in \mathcal{L}(X_{G_{NS}})$ *and* K *is a compact operator.*

Proof. The desired space is $X_{G_{NS}} = T\left[G_{NS}, \left(\mathcal{S}_{n_j}, \frac{1}{m_j}\right)_j, D\right]$.

Indeed (1),(2),(3) and (7) follow from Theorem 9.8. Also (4),(6) have been shown in Proposition 10.11 since the set $\Gamma = \mathcal{B}$ of all branches of the binary tree has the cardinality of the continuum. (5) has been shown in Theorem 10.12 and Proposition 10.14. Finally Proposition 1.7 yields (8). □

THEOREM 10.16. *There exists a nonseparable Banach space* Z, *which is a dual of a separable Banach space, such that every* $T \in \mathcal{L}(Z)$ *is of the form* $T = \lambda I + R$ *where* $\lambda \in \mathbb{R}$ *and* R *is a weakly compact operator with separable range.*

Proof. For the space $X_{G_{NS}} = T\left[G_{NS}, \left(\mathcal{S}_{n_j}, \frac{1}{m_j}\right)_j, D\right]$, the nonseparable space $Z = X_{G_{NS}}^*$ satisfies the conclusion of the theorem. Indeed, by Theorem 10.15(8), every $T \in \mathcal{L}(Z) = \mathcal{L}(X_{G_{NS}}^*)$ is of the form $T = Q^* + K$ where $Q \in \mathcal{L}(X_{G_{NS}})$ and K is a compact operator. Also (7) of Theorem 10.15 yields that $Q = \lambda I + L$ for some weakly compact operator L and a $\lambda \in \mathbb{R}$. The operator $R = L^* + K$ is weakly compact, and $T = \lambda I + R$. The set $\overline{RB_{X_{G_{NS}}^*}}$ is norm separable, as it is a weakly compact set in the dual of a separable Banach space. Therefore $\overline{RX_{G_{NS}}^*}$ is separable. □

REMARK 10.17. The existence of a nonseparable Banach space X such that every $T \in \mathcal{L}(X)$ is of the form $T = \lambda I + R$ with R operator with separable range was established by S. Shelah in [**48**] assuming that $V = L$. Subsequently S. Shelah and J. Steprans in [**49**] proved the existence of such a space without any additional set-theoretical assumption.

CHAPTER 11

Complemented embedding of ℓ^p, $1 \leq p < \infty$, in the duals of H.I. spaces

For $1 < p < \infty$, two spaces $T\big[G_p, \Big(\mathcal{S}_{n_j}, \frac{1}{m_j}\Big)_j, D\big]$ and $T\big[G_p^{\mathcal{L}}, \Big(\mathcal{S}_{n_j}, \frac{1}{m_j}\Big)_j, D\big]$ are presented in this chapter. Both are related to the space ℓ^p in the sense that ℓ^p is a quotient of each of them and ℓ^q, $\frac{1}{p} + \frac{1}{q} = 1$ is w^*-isometric to a complemented subspace of the duals of the spaces. These spaces are asymptotic ℓ^1 and H.I.

Let $1 < p < \infty$ and q such that $\frac{1}{p} + \frac{1}{q} = 1$. We set

$$G_p = \Big\{\sum_{i=1}^{d} a_i s_i^* : \ (a_i)_{i=1}^d \in \mathbb{Q}^d \text{ with } \sum_{i=1}^{d} |a_i|^q \leq 1 \text{ and } (s_i)_{i=1}^d \text{ are pairwise disjoint finite segments of } \mathcal{D},\ d \in \mathbb{N}\Big\}$$

The set G_p is ground (Definition 3.1) while the closure of G_p in the topology of pointwise convergence is the set

$$\overline{G_p}^p = \Big\{\sum_{i=1}^{\infty} a_i s_i^* : \ (a_i)_{i\in\mathbb{N}} \text{ with } \sum_{i=1}^{\infty} |a_i|^q \leq 1 \text{ and } (s_i)_{i=1}^\infty \text{ are pairwise disjoint segments of } \mathcal{D}\Big\}$$

Note that in the case $p = 2$ the ground space Y_{G_2}, which by definition is the the completion of the space $(c_{00}, \|\ \|_{G_2})$, is exactly James tree space [**33**] induced by R.C. James, the first example of a separable space not containing ℓ^1 that has nonseparable dual.

We recall that a Banach space Z satisfies the weak Banach-Saks property if every weakly null sequence in Z has a norm Cesaro summable subsequence.

LEMMA 11.1. Let G be a ground set such that ℓ^1 does not embed into the ground space Y_G. If Y_G satisfies the weak Banach-Saks property then the set G is $\mathcal{S}_2$ bounded.

Proof. Let $(y_n)_{n\in\mathbb{N}}$ be a weakly null block sequence in Y_G. Since the space Y_G satisfies the weak Banach-Saks property we may assume, passing to a subsequence, that the sequence $(y_n)_{n\in\mathbb{N}}$ is norm Cesaro summable. By standard arguments (e.g. [**39**], [**13**]) it is shown that for every $\varepsilon > 0$ there exists $k_\varepsilon \in \mathbb{N}$ and $L \in [\mathbb{N}]$ such that for all $g \in G$, $\#\{n \in L : \ |g(y_n)| > \varepsilon\} \leq k_\varepsilon$. Hence if we assume that $\min L \geq k_\varepsilon$ we obtain that $\{\min\operatorname{supp} y_n : \ n \in L,\ |g(y_n)| > \varepsilon\} \in \mathcal{S}_1 \subset \mathcal{S}_2$. Hence the ground set G is $\mathcal{S}_2$ bounded. $\square$

LEMMA 11.2. The set G_p is $\mathcal{S}_2$ bounded.

Proof. A result of I. Amemiya and T. Ito (the proof given in [**2**] for $p = 2$ clearly remains valid for any $1 < p < \infty$) implies that every weakly null sequence in Y_{G_p} has a subsequence isomorphic to the standard basis of ℓ^p; this subsequence is norm Cesaro summable. Lemma 11.1 yields that the set G_p is $\mathcal{S}_2$ bounded. $\square$

The space X_{G_p} is defined as

$$X_{G_p} = T\left[G_p, \left(\mathcal{S}_{n_j}, \frac{1}{m_j}\right)_j, D\right].$$

LEMMA 11.3. For $b_1, b_2, \ldots, b_d$ pairwise disjoint infinite segments of $\mathcal{D}$ and $a_1, a_2, \ldots, a_d \in \mathbb{R}$ we have that

$$\|\sum_{i=1}^{d} a_i b_i^* \|_{X^*_{G_p}} = (\sum_{i=1}^{d} |a_i|^q)^{\frac{1}{q}}$$

Proof. Assuming that $\sum\limits_{i=1}^{d} |a_i|^q = 1$ we'll show that $\| \sum\limits_{i=1}^{d} a_i b_i^* \| = 1$. The inequality $\| \sum\limits_{i=1}^{d} a_i b_i^* \| \leq 1$ is straightforward since $\sum\limits_{i=1}^{d} a_i b_i^* \in \overline{G_p}^{w^*} \subset B_{X^*_{G_p}}$.

It suffices to show that $\| \sum\limits_{i=1}^{d} a_i b_i^* \| \geq \dfrac{1}{1+\varepsilon}$ for every $\varepsilon > 0$. Let $\varepsilon > 0$ and choose $j \in \mathbb{N}$ such that

$$\frac{1}{m_j} < \frac{\varepsilon}{2d}$$

Since $b_1, b_2, \ldots, b_d$ are pairwise different we may select a $k_0 \in \mathbb{N}$ such that $b_1 \cap E_{k_0}, b_2 \cap E_{k_0}, \ldots, b_n \cap E_{k_0}$ are incomparable segments of $\mathcal{D}$, where $E_{k_0} = \{n \in \mathbb{N} : n \geq 2^{k_0}\}$.

Choose $L^1, L^2, \ldots, L^d$ infinite subsets of $\mathbb{N}$, $L^i = (l_t^i)_{t \in \mathbb{N}}$ $i = 1, 2, \ldots, d$, satisfying the following conditions:

(i) $L^i \subset b_i \cap E_{k_0}$ for $i = 1, 2, \ldots d$.
(ii) $l_t^1 < l_t^2 < \cdots < l_t^d < l_{t+1}^1 < \cdots$ for each $t = 1, 2, \ldots$
(iii) $l_1^1 > \dfrac{6d}{\varepsilon}$.

Let $a(n_j, L^1) = \sum\limits_{t=1}^{n} r_t e_{l_t^1}$ and set

$$x_i = \sum_{t=1}^{n} r_t e_{l_t^i} \in X_{G_p} \quad \text{and} \quad \tilde{a}_i = \operatorname{sgn}(a_i) \cdot |a_i|^{\frac{q}{p}} \text{ for } i = 1, 2, \ldots, d.$$

First observe that $(\sum\limits_{i=1}^{d} a_i b_i)(\sum\limits_{i=1}^{d} \tilde{a}_i x_i) = \sum\limits_{i=1}^{d} a_i \tilde{a}_i = \sum\limits_{i=1}^{d} |a_i|^q = 1$.

To finish the proof it it enough to show that $\| \sum\limits_{i=1}^{d} \tilde{a}_i x_i \| \leq 1 + \varepsilon$, since then

$$\|\sum a_i b_i^*\| \geq \frac{(\sum\limits_{i=1}^{d} a_i b_i)(\sum\limits_{i=1}^{d} \tilde{a}_i x_i)}{\|\sum\limits_{i=1}^{d} \tilde{a}_i x_i\|} \geq \frac{1}{1+\varepsilon}.$$

Since the norming set D of the space $X_{G_p} = T\left[G_p, \left(\mathcal{S}_{n_j}, \frac{1}{m_j}\right)_j, D\right]$ is a subset of the set $W_p = W[G_p, \left(\mathcal{S}_{n_j}, \frac{1}{m_j}\right)_j]$ it suffices to show that $f(\sum_{i=1}^{d} \tilde{a}_i x_i) \leq 1 + \varepsilon$ for every $f \in A \cap W_p$. Let $f \in A \cap W_p$. Using similar arguments as in Definition 3.14 and Lemma 3.15 we may assume that there exists a tree $T_f = (f_a)_{a \in \mathcal{A}}$ of f (Definition 3.5) such that each f_a is not of type II (Definition 3.6). Let $(g_{a_s})_{s=1}^{s_0}$ be the functionals corresponding to the maximal elements of the tree $\mathcal{A}$. We denote by $\preceq$ the ordering of the tree $\mathcal{A}$. Let

$$A = \left\{ s \in \{1, 2, \ldots, s_0\} : \prod_{\gamma \prec a_s} \frac{1}{w(f_\gamma)} \leq \frac{1}{m_j} \right\}$$

$$B = \{1, 2, \ldots, s_0\} \setminus A$$

and set $f_A = f|_{\bigcup_{s \in A} \operatorname{supp} g_{a_s}}$, $f_B = f|_{\bigcup_{s \in B} \operatorname{supp} g_{a_s}}$.

We have $f_A(x_i) \leq \frac{1}{m_j}$ for each i thus

$$f_A(\sum_{i=1}^{d} \tilde{a}_i x_i) \leq \frac{1}{m_j} \sum_{i=1}^{d} |\tilde{a}_i| \leq \frac{1}{m_j} \cdot d < \frac{\varepsilon}{2}. \tag{35}$$

It remains to estimate the value $f_B(\sum_{i=1}^{d} \tilde{a}_i x_i)$. We observe that

$$\sum_{i=1}^{d} \tilde{a}_i x_i = \sum_{i=1}^{d} \tilde{a}_i (\sum_{t=1}^{n} r_t e_{l_t^i}) = \sum_{t=1}^{n} r_t (\sum_{i=1}^{d} \tilde{a}_i e_{l_t^i}).$$

Set

$$\begin{aligned} E_1 &= \Big\{ t \in \{1, 2, \ldots, n\} : \text{ the set } \{l_t^1, l_t^2, \ldots, l_t^d\} \text{ is contained in } \operatorname{ran} g_{a_s} \\ &\qquad \text{for some } s \in B \text{ or does not intersect any } \operatorname{ran} g_{a_s},\ s \in B \Big\} \\ E_2 &= \{1, 2, \ldots, n\} \setminus E_1. \end{aligned}$$

For each $s = 1, 2, \ldots, s_0$ set $\theta_s = \sum \Big\{ r_t : \ \{l_t^1, l_t^2, \ldots, l_t^d\} \subset \operatorname{ran} g_{a_s} \Big\}$ and observe that $\sum_{s \in B} \theta_s \leq 1$.

We first estimate the quantity $g_{a_s}(\sum_{t \in E_1} r_t(\sum_{i=1}^{d} \tilde{a}_i e_{l_t^i}))$ for $s \in B$. We may assume that g_{a_s} is of the form $g_{a_s} = \sum_i \sum_j c_{i,j} s_{i,j}^*$ where $\{s_{i,j} : i, j\}$ is a family of pairwise disjoint segments with each $s_{i,j} \subset b_i$ and $\sum_i \sum_j |c_{i,j}|^q \leq 1$. For $1 \leq i' \leq d$ we get that $(\sum_j c_{i',j} s_{i',j}^*)(\sum_{t \in E_1} r_t(\sum_{i=1}^{d} \tilde{a}_i e_{l_t^i})) = \tilde{a}_{i'}(\sum_j c_{i',j} s_{i',j}^*)(\sum_{t \in E_1} r_t e_{l_t^{i'}}) \leq |\tilde{a}_{i'}| \max_j |c_{i',j}| \theta_s$.
Thus

$$g_{a_s}(\sum_{t \in E_1} r_t(\sum_{i=1}^{d} \tilde{a}_i e_{l_t^i})) \leq \theta_s \sum_{i=1}^{d} |\tilde{a}_i| \max_j |c_{i,j}| \leq \theta_s (\sum_{i=1}^{d} \max_j |c_{i,j}|^q)^{\frac{1}{q}} (\sum_{i=1}^{d} |\tilde{a}_i|^p)^{\frac{1}{p}} \leq \theta_s.$$

Therefore

$$f_B\Big(\sum_{t\in E_1} r_t\Big(\sum_{i=1}^d \tilde{a}_i e_{l_t^i}\Big)\Big) \le \Big(\sum_{s\in B} g_{a_s}\Big)\Big(\sum_{t\in E_1} r_t\Big(\sum_{i=1}^d \tilde{a}_i e_{l_t^i}\Big)\Big) \le \sum_{s\in B}\theta_s \le 1. \tag{36}$$

By the definition of the set B we may prove, as in the proof of Lemma 3.16, that the family of functionals $\{g_{a_s} : s \in B\}$ is $\mathcal{S}_{(\log_2(m_j)-1)n_{j-1}}$ admissible. By the definition of the set E_2 for $t \in E_2$ the set $\{l_t^1, l_t^2, \dots, l_t^d\}$ intersects at least one but is not contained in any $\operatorname{ran} g_{a_s}$, $s \in B$. These easily yield that $\{l_t^1 : t \in E_2\} \in \mathcal{S}_{\log_2(m_j)n_{j-1}} \subset \mathcal{S}_{n_j-1}$ and thus $\sum_{t\in E_2} r_t < \frac{3}{l_1^1}$. Therefore

$$f_B\Big(\sum_{t\in E_2} r_t\Big(\sum_{i=1}^d \tilde{a}_i e_{l_t^i}\Big)\Big) \le \Big(\sum_{t\in E_2} r_t\Big)\Big(\sum_{i=1}^d |\tilde{a}_i|\Big) < \frac{3}{l_1^1}\cdot d < \frac{\varepsilon}{2}. \tag{37}$$

From (35),(36) and (37), we conclude that

$$\begin{aligned} f\Big(\sum_{i=1}^d \tilde{a}_i x_i\Big) &\le f_A\Big(\sum_{i=1}^d \tilde{a}_i x_i\Big) + f_B\Big(\sum_{t\in E_1} r_t\Big(\sum_{i=1}^d \tilde{a}_i e_{l_t^i}\Big)\Big) + f_B\Big(\sum_{t\in E_2} r_t\Big(\sum_{i=1}^d \tilde{a}_i e_{l_t^i}\Big)\Big) \\ &\le \frac{\varepsilon}{2} + 1 + \frac{\varepsilon}{2} = 1 + \varepsilon. \end{aligned}$$

□

PROPOSITION 11.4. (i) If $(s_i)_{i\in\mathbb{N}}$ is any sequence of pairwise disjoint infinite segments of $\mathcal{D}$ then $\overline{\operatorname{span}}\{s_i^* : i \in \mathbb{N}\}$ is isometric to ℓ^q.

(ii) The quotient space $X_{G_p}^*/(X_{G_p})_*$ is isometric to $\ell^q(\mathcal{B})$.

(iii) If $Q : X_{G_p}^* \longrightarrow X_{G_p}^*/(X_{G_p})_*$ is the quotient map then for every countable subset $\mathcal{C}$ of the set of branches $\mathcal{B}$ there exists a subspace $Z_{\mathcal{C}}$ of $X_{G_p}^*$ such that Q maps isometrically $Z_{\mathcal{C}}$ to $\ell^q(\mathcal{C})$.

Proof. The assertion (i) follows readily from the previous lemma. To prove (ii) let $(b_i)_{i=1}^n$ be pairwise different branches. Let k_0 be an integer such that the infinite segments $(b_i \cap E_{k_0})_{i=1}^n$ are pairwise disjoint. The same holds for all $k \ge k_0$. For every choice $(\lambda_i)_{i=1}^n$ of scalars we have that

$$\begin{aligned} &\|\lambda_1(b_1^* + (X_{G_p})_*) + \cdots + \lambda_n(b_n^* + (X_{G_p})_*)\| \\ &= \operatorname{dist}(\lambda_1 b_1^* + \cdots + \lambda_n b_n^*,\ (X_{G_p})_*) \\ &= \operatorname{dist}(\lambda_1 b_1^* + \cdots + \lambda_n b_n^*,\ \overline{\operatorname{span}}\{e_j^* : j \in \mathbb{N}\}) \\ &= \operatorname{dist}(\lambda_1 b_1^* + \cdots + \lambda_n b_n^*,\ \operatorname{span}\{e_j^* : j \in \mathbb{N}\}) \\ &= \lim_k \operatorname{dist}(\lambda_1 b_1^* + \cdots + \lambda_n b_n^*,\ \operatorname{span}\{e_j^* : j \le 2^k - 1\}) \\ &= \lim_k \|\lambda_1(b_1 \cap E_k)^* + \cdots + \lambda_n(b_n \cap E_k)^*\| \\ &= \Big(\sum_{i=1}^n |\lambda_i|^q\Big)^{\frac{1}{q}} \end{aligned}$$

Let $\mathcal{C} = \{b_i : i \in \mathbb{N}\}$ be a countable set of branches. We may inductively obtain a sequence $(s_i)_{i\in\mathbb{N}}$ of pairwise disjoint infinite segments such that $s_i \subset b_i$ for all i in the following manner: Set $s_1 = b_1$. Suppose that $s_1, s_2, \dots, s_i$ have been defined. Choose k_{i+1} such that $b_{i+1} \cap E_{k_{i+1}}$ does not intersect $s_1, s_2, \dots, s_i$ and set $s_{i+1} = b_{i+1} \cap E_{k_{i+1}}$.

The space $Z_{\mathcal{C}} = \overline{\text{span}}\{s_i^* : i \in \mathbb{N}\}$ is mapped isometrically to $\ell^q(\mathcal{C})$ by the quotient map Q. Therefore ℓ^q is embedded into $X_{G_p}^*$ as a complemented subspace. □

DEFINITION 11.5. Let $(\Lambda_i)_{i\in\mathbb{N}}$ be any partition of $\mathbb{N}$ into infinite sets, $\Lambda_i = \{\lambda_1^i < \lambda_2^i < \cdots\}$ i=1,2,... This partition naturally induces a tree structure $(\mathcal{L}, \preceq)$ whose set of nodes is $\mathcal{L} = \mathbb{N}$ and $\preceq$ is defined as follows:

$$n \preceq m \text{ iff } n \leq m \text{ and there exists } i \in \mathbb{N} \text{ such that } n, m \in \Lambda_i$$

A finite segment of $\mathcal{L}$ is of the form $\{\lambda_t^i : k_1 \leq t \leq k_2\}$ for some $i, k_1, k_2 \in \mathbb{N}$, while infinite segments are of the form $\{\lambda_t^i : t \geq k_1\}$ for some $i, k_1 \in \mathbb{N}$. Branches of $\mathcal{L}$ are exactly the infinite sets Λ_i, i=1,2,... of the partition of $\mathbb{N}$ which defined the tree $\mathcal{L}$. Note that the set of all segments of $\mathcal{L}$ is countable.

THEOREM 11.6. *For $1 < p < \infty$ there exists an asymptotic ℓ^1 H.I. Banach space Y_p with a Schauder basis such that:*

(i) *The space ℓ^p is a quotient space of Y_p.*
(ii) *The dual space Y_p^* is isomorphic to $(Y_p)_* \oplus \ell^q$ and $(Y_p)_*$ is a H.I. space.*

Proof. Let $G_p^{\mathcal{L}}$ be the set

$$G_p^{\mathcal{L}} = \Big\{\sum_{i=1}^{d} a_i s_i^* : (s_i)_{i=1}^d \text{ are finite segments of } \mathcal{L},$$
$$s_i \subset \Lambda_i,\ a_i \in \mathbb{Q} \text{ for } i = 1, 2, \ldots, d \text{ and } \sum_{i=1}^{d} |a_i|^q \leq 1,\ d \in \mathbb{N}\Big\}$$

As in Lemma 11.2 we can prove that the set $G_p^{\mathcal{L}}$ is $\mathcal{S}_2$ bounded. The space Y_p is the space $Y_p = X_{G_p^{\mathcal{L}}} = T\left[G_p^{\mathcal{L}}, \left(\mathcal{S}_{n_j}, \frac{1}{m_j}\right)_j, D\right]$ and the desired properties are proved using similar arguments to those of Proposition 11.4. □

THEOREM 11.7. *There exists an asymptotic ℓ^1 Banach space X with a boundedly complete basis $(e_n)_{n\in\mathbb{N}}$ such that*

(i) *X is H.I. and has a H.I. predual X_*.*
(ii) *X^* is separable and H.I.*
(iii) *X^*/X_* is isometric to $c_0(\mathbb{N})$.*
(iv) *X^{**} is isomorphic to $X \oplus \ell^1(\mathbb{N})$.*

Proof. We set

$$G_0^{\mathcal{L}} = \Big\{\sum_{i=1}^{d} \varepsilon_i s_i^* : (s_i)_{i=1}^d \text{ finite segments of } \mathcal{L}, s_i \subset \Lambda_i,\ (\varepsilon_i)_{i=1}^d \in \{-1,1\}^d$$
$$\{\min s_1, \min s_2, \ldots, \min s_d\} \in \mathcal{S},\ d \in \mathbb{N}\Big\}.$$

Then the space $X_{G_0^{\mathcal{L}}} = T[G_0^{\mathcal{L}}, (\mathcal{S}_{n_j}, \frac{1}{m_j})_j, D]$ satisfies the desired properties (i),(ii),(iii) and (iv). Thus $X_{G_0^{\mathcal{L}}}^*$ is an example of a separable H.I. Banach space, such that $\ell^1(\mathbb{N})$ is a complemented subspace of its dual. □

CHAPTER 12

Compact families in $\mathbb{N}$

As we have mentioned in the introduction the spaces $T[G,(\mathcal{S}_{\xi_j},\frac{1}{m_j})_j,D]$ are defined in a similar manner to the spaces $T[G,(\mathcal{S}_{n_j},\frac{1}{m_j})_j,D]$. Moreover the results proved for the general case are analogous to what we have proved in the previous chapters. However there are some difficulties when we deal with $(\mathcal{S}_\xi)_{\xi<\omega_1}$ instead of $(\mathcal{S}_n)_{n<\omega}$. Indeed, while for $n<\omega$ the modified version $\mathcal{S}_n^M$ of the $n-$th Schreier family coincides with $\mathcal{S}_n$ ([**7**]) for general $\xi<\omega_1$ we are not able to prove the same. Also for $n<\omega$ $\mathcal{S}_n*\mathcal{S}_n=\mathcal{S}_{2n}$ and again, for general $\xi<\omega_1$, it is not clear the relation between $\mathcal{S}_\xi*\mathcal{S}_\xi$ and $\mathcal{S}_{\xi\cdot 2}$. Thus in the present chapter we study the convolution between compact families of finite subsets of $\mathbb{N}$ and we prove some inclusion results for $\mathcal{S}_{\xi_1}*\mathcal{S}_{\xi_2}*\cdots*\mathcal{S}_{\xi_n}$ and $\mathcal{S}_\xi^M$. These results are used in the next chapter.

The Cantor-Bendixson derivative of a hereditary and compact family $\mathcal{M}$ of finite subsets of $\mathbb{N}$ is defined as

$$\mathcal{M}'=\{F\in\mathcal{M}:\ \text{there exists } L\in[\mathbb{N}] \text{ such that } F\cup\{l\}\in\mathcal{M} \text{ for every } l\in L\}$$

The ξth derivative of a family $\mathcal{M}$, denoted by $\mathcal{M}^{(\xi)}$, is defined recursively as follows:

$$\mathcal{M}^{(1)}=\mathcal{M}'$$

$$\mathcal{M}^{(\xi+1)}=(\mathcal{M}^{(\xi)})'$$

If ξ is a limit ordinal and $\mathcal{M}^{(\zeta)}$ has been defined for all $\zeta<\xi$ we set

$$\mathcal{M}^{(\xi)}=\bigcap_{\zeta<\xi}\mathcal{M}^{(\zeta)}.$$

The Cantor-Bendixson index of $\mathcal{M}$, denoted by $i(\mathcal{M})$, is the least countable ordinal ξ_0 such that $\mathcal{M}^{(\xi_0)}\subset\{\emptyset\}$. (This is a slight variation of the original definition of the Cantor-Bendixson index for a countable compact metric space which is defined as the least ordinal ξ_0 that satisfies $\mathcal{M}^{(\xi_0)}=\emptyset$.)
It is proved in [**1**] that the Cantor-Bendixson index of the Schreier family $\mathcal{S}_\xi$ is $i(\mathcal{S}_\xi)=\omega^\xi$.

We recall at this point that if $\mathcal{M}$ and $\mathcal{N}$ are two families of finite subsets of $\mathbb{N}$, the convolution of $\mathcal{M}$ and $\mathcal{N}$ is defined as

$$\begin{aligned}\mathcal{M}*\mathcal{N} \;=\; &\Big\{F\subset\mathbb{N}:\ \text{there exists } s\in\mathbb{N},\ F_1,F_2,\dots,F_s\in\mathcal{N} \text{ and}\\ &\{m_1<m_2<\cdots<m_s\}\in\mathcal{M} \text{ such that } m_1\le F_1<m_2\le F_2<\\ &\cdots<m_s\le F_s \text{ and } F=\bigcup_{i=1}^s F_i\Big\}\cup\Big\{\emptyset\Big\}.\end{aligned}$$

PROPOSITION 12.1. Let $\mathcal{M}, \mathcal{N}$ be a pair of hereditary compact and spreading families of finite sets of $\mathbb{N}$. If $i(\mathcal{M}) = \xi$ and $i(\mathcal{N}) = \zeta$ then

$$i(\mathcal{M} * \mathcal{N}) \leq \zeta \cdot \xi.$$

Proof. It clearly suffices to show the following statement:
If $\mathcal{N}$ is a hereditary compact and spreading family of finite subsets of $\mathbb{N}$ with $i(\mathcal{N}) = \zeta$ and $a \geq 1$ is a countable ordinal then for every hereditary compact and spreading family $\mathcal{M}$ we have that

$$(\mathcal{M} * \mathcal{N})^{(\zeta \cdot a)} \subset \mathcal{M}^{(a)} * \mathcal{N}.$$

We proceed with the first inductive step, namely for $a = 1$. Assume the contrary. Then there exists $F \in (\mathcal{M} * \mathcal{N})^{(\zeta)}$ such that $F \notin \mathcal{M}^{(1)} * \mathcal{N}$. We can write

$$F = F_1 \cup F_2 \cup \cdots \cup F_{d-1} \cup F_d$$

with $F_1, F_2, \ldots, F_{d-1}$ successive maximal sets of $\mathcal{N}$, $F_d \neq \emptyset$ and $\{m_1, m_2, \ldots, m_d\} \in \mathcal{M} \setminus \mathcal{M}^{(1)}$ where $m_i = \min F_i$ for $i = 1, 2, \ldots, d$.

The $\mathcal{N}$ maximality of the sets F_i, $i = 1, 2, \ldots, d-1$ yields that for every finite set G such that $F < G$ and $F \cup G \in \mathcal{M} * \mathcal{N}$ we have $F_d \cup G \in \mathcal{N}$. It follows that F_d is a nonempty member of $\mathcal{N}^{(\zeta)}$, a contradiction since $i(\mathcal{N}) = \zeta$.

Let a be a countable ordinal and suppose that $(\mathcal{M} * \mathcal{N})^{(\zeta \cdot \beta)} \subset \mathcal{M}^{(\beta)} * \mathcal{N}$ for all $\beta < a$.
If $a = \beta + 1$ then

$$\begin{aligned} (\mathcal{M} * \mathcal{N})^{(\zeta \cdot a)} &= (\mathcal{M} * \mathcal{N})^{(\zeta \cdot \beta + \zeta)} = ((\mathcal{M} * \mathcal{N})^{(\zeta \cdot \beta)})^{(\zeta)} \\ &\subset (\mathcal{M}^{(\beta)} * \mathcal{N})^{(\zeta)} \subset (\mathcal{M}^{(\beta)})^{(1)} * \mathcal{N} \\ &= \mathcal{M}^{(a)} * \mathcal{N} \end{aligned}$$

If a is a limit ordinal select a strictly increasing sequence $(a_n)_{n \in \mathbb{N}}$ of countable ordinals with $\sup a_n = a$. Let $F \in (\mathcal{M} * \mathcal{N})^{(\zeta \cdot a)}$. Then $F \in (\mathcal{M} * \mathcal{N})^{(\zeta \cdot a_n)}$ for all n and thus by the inductive assumption $F \in \mathcal{M}^{(a_n)} * \mathcal{N}$ for all n. For each n let $M_n = \{m_1^n, m_2^n, \ldots, m_{d_n}^n\} \in \mathcal{M}^{(a_n)}$ and $F_1^n, F_2^n, \ldots, F_{d_n}^n$ be nonempty sets of $\mathcal{N}$ such that $m_1^n \leq F_1^n < m_2^n \leq F_2^n < \ldots \leq m_{d_n}^n < F_{d_n}^n$ and $F = \bigcup_{i=1}^{d_n} F_i^n$. Since each M_n is subset of the finite set $\{1, 2, \ldots, \max F\}$ we can find a stable subsequence $(M_{k_n})_{n \in \mathbb{N}}$ let $M_{k_n} = M$ for all n. We thus get $M \in \bigcap_{n \in \mathbb{N}} \mathcal{M}^{(a_{k_n})} = \mathcal{M}^{(a)}$ which yields that $F \in \mathcal{M}^{(a)} * \mathcal{N}$.

This completes the proof of the inductive assumption and the entire proof of the proposition. □

LEMMA 12.2. If $\mathcal{M}$ is a compact and spreading family of finite subsets of $\mathbb{N}$ and $L \in [\mathbb{N}]$, $L = \{l_1 < l_2 < \cdots\}$ then the families $\mathcal{M}$, $\mathcal{M}[L] = \{F \in \mathcal{M} : F \subset L\}$ and $\mathcal{M}(L) = \{L(F) : F \in \mathcal{M}\}$ (where $L(F) = \{l_i : i \in F\}$) have the same Cantor-Bendixson index.

Proof. Since $\mathcal{M}$ is spreading we have $\mathcal{M}(L) \subset \mathcal{M}[L]$ and thus $\mathcal{M}(L) \subset \mathcal{M}[L] \subset \mathcal{M}$. Since also $\mathcal{M}(L)$ is naturally homeomorphic to $\mathcal{M}$ it follows from a theorem of Mazurkiewicz-Sierpinski [**38**] that $\mathcal{M}[L]$ is also homeomorphic to $\mathcal{M}$. Therefore the families $\mathcal{M}, \mathcal{M}[L]$ and $\mathcal{M}(L)$ have the same Cantor-Bendixson index. □

The following dichotomy is obtained in [**24**].

THEOREM 12.3. *Let $\mathcal{M}, \mathcal{N}$ be a pair of hereditary families of finite subsets of $\mathbb{N}$ and $P \in [\mathbb{N}]$. Then there exists $L \in [P]$ such that either $\mathcal{M}[L] \subset \mathcal{N}$ or $\mathcal{N}[L] \subset \mathcal{M}$.*

By Lemma 12.2 and Theorem 12.3 we get the following Corollary:

COROLLARY 12.4. Let $\mathcal{M}, \mathcal{N}$ be two hereditary compact families of finite subsets of $\mathbb{N}$ such that $\mathcal{N}$ is spreading and $i(\mathcal{M}) < i(\mathcal{N})$. Then for every $P \in [\mathbb{N}]$ there exists an $L \in [P]$ such that $\mathcal{M}[L] \subset \mathcal{N}$.

We give the definition of the repeated averages hierarchy introduced in [**13**]. Let $(e_n)_{n\in\mathbb{N}}$ denote the standard basis of c_{00}. For every $\xi < \omega_1$ and $L \in [\mathbb{N}]$ a convex block sequence $(\xi_n^L)_{n\in\mathbb{N}}$ of $(e_n)_{n\in\mathbb{N}}$ is defined. The definition is recursive on ξ in the following manner:
For $\xi = 0$, we set $\xi_n^L = e_{l_n}$ where $L = \{l_1 < l_2 < l_3 < \cdots\}$.
Let $0 < \xi < \omega_1$ and suppose that the sequence $(\zeta_n^L)_{n\in\mathbb{N}}$ has been defined for every $\zeta < \xi$ and $L \in [\mathbb{N}]$.
If $\xi = \zeta + 1$ then

$$\xi_1^L = \frac{1}{l_1}\sum_{i=1}^{l_1} \zeta_i^L$$

where $l_1 = \min L$. Suppose that $\xi_1^L < \xi_2^L < \cdots < \xi_n^L$ have been defined. Let $L_n = \{l \in L : l > \max\operatorname{supp}\xi_n^L\}$ and $k_n = \min L_n$.

$$\xi_{n+1}^L = \frac{1}{k_n}\sum_{i=1}^{k_n} \zeta_i^{L_n} (= \xi_1^{L_n})$$

If ξ is a limit ordinal and $(\xi_n)_{n\in\mathbb{N}}$ is the increasing sequence of non limit ordinals with $\sup \xi_n = \xi$ used in the definition of the family $\mathcal{S}_\xi$, the sequence $(\xi_n^L)_{n\in\mathbb{N}}$ for $L = \{l_1 < l_2 < l_3 < \cdots\}$ is defined as follows:

$$\xi_1^L = [\xi_{l_1}]_1^L$$

Suppose that $\xi_1^L < \xi_2^L < \cdots < \xi_n^L$ have been defined. Let $L_n = \{l \in L : l > \max\operatorname{supp}\xi_n^L\}$ and $k_n = \min L_n$. Then

$$\xi_{n+1}^L = [\xi_{k_n}]_1^{L_n}$$

The following properties of the hierarchies $\{\mathcal{S}_\xi : \xi < \omega_1\}$ and $\{\xi_n^L : L \in [\mathbb{N}],\ n \in \mathbb{N},\ \xi < \omega_1\}$ have been established in [**13**].

P.1 For every $\xi < \omega_1$ and $L \in [\mathbb{N}]$, $(\xi_n^L)_{n\in\mathbb{N}}$ is a block sequence of elements of $S_{\ell^1}^+$ and $L = \bigcup_{n=1}^{\infty} \operatorname{supp}\xi_n^L$.

P.2 For every $\xi < \omega_1$, $L \in [\mathbb{N}]$ and $n \in \mathbb{N}$, $\operatorname{supp}\xi_n^L \in \mathcal{S}_\xi$. If for every limit ordinal $\xi < \omega_1$ the sequence $(\xi_n)_{n\in\mathbb{N}}$ has been defined such that the sequence $(\mathcal{S}_{\xi_n})_{n\in\mathbb{N}}$ is increasing then $\operatorname{supp}\xi_n^L$ is a maximal $\mathcal{S}_\xi$ set for every $\xi < \omega_1$, $L \in [\mathbb{N}]$ and $n \in \mathbb{N}$.

P.3 For every $\xi < \omega_1$ and every $L_1, L_2 \in [\mathbb{N}]$ such that $\operatorname{supp}\xi_i^{L_1} = \operatorname{supp}\xi_i^{L_2}$ for $i = 1, 2, \ldots, k$ we have that $\xi_i^{L_1} = \xi_i^{L_2}$ for $i = 1, 2, \ldots, k$.

P.4 For every $\xi < \omega_1$, if $(t_i)_{i\in\mathbb{N}}$ is a sequence in $\mathbb{N}$ and $(L_i)_{i\in\mathbb{N}}$ is a sequence in $[\mathbb{N}]$ such that $\operatorname{supp}\xi_{t_i}^{L_i} < \operatorname{supp}\xi_{t_{i+1}}^{L_{i+1}}$ for all i then if $L = \bigcup_{i=1}^{\infty} \operatorname{supp}\xi_{t_i}^{L_i}$ we have that $\xi_i^L = \xi_{t_i}^{L_i}$ for each $i \in \mathbb{N}$.

P.5 If $\operatorname{supp}\xi_n^L = \{t_1 < t_2 < \cdots < t_d\}$ and $\xi_n^L = \sum_{i=1}^{d} a_{t_i} e_{t_i}$ then $a_{t_1} \geq a_{t_2} \geq \cdots \geq a_{t_d}$.

The properties **P.3** and **P.4** are called stability properties of the repeated averages hierarchy.

NOTATION 12.5. 1. For a sequence $\vec{x} = (x_n)_{n\in\mathbb{N}}$ in a Banach space and a vector $\tau = \sum a_n e_n$ in c_{00} we denote by $\tau \cdot \vec{x}$ the vector $\sum_{n=1}^{\infty} a_n x_n$.
2. If $\vec{x} = (x_n)_{n\in\mathbb{N}}$ is a block sequence in c_{00} and $\xi < \omega_1$ we use the notation

$$a(\xi, \vec{x}) = a(\xi, (x_n)_{n\in\mathbb{N}}) = \sum_k \xi_1^L(t_k) x_k.$$

where $L = \{t_k : k \in \mathbb{N}\}$ and $t_k = \min \operatorname{supp} x_k$ for all k.

LEMMA 12.6. The Cantor-Bendixson index of the modified Shreier family $\mathcal{S}_\xi^M$ is $i(\mathcal{S}_\xi^M) = \omega^\xi$.

Proof. We proceed by induction. The case $\xi = 0$ is trivial. Let $\xi < \omega_1$ and suppose that $i(\mathcal{S}_\zeta^M) = \omega^\zeta$ for all $\zeta < \xi$.

If $\xi = \zeta + 1$ using the same methods as in [**1**], we may prove that for every $n < \omega$

$$(\mathcal{S}_{\zeta+1}^M)^{(\omega^\zeta \cdot n)} = \Big\{F \subset \mathbb{N} : \ \exists k \in \mathbb{N} \text{ and } F_1, F_2, \dots, F_k \in \mathcal{S}_\zeta^M \text{ such that } k + n \leq F_i \ \ \forall i = 1, 2, \dots, k \text{ and } F = \bigcup_{i=1}^{k} F_i\Big\}$$

and we easily deduce that $i(\mathcal{S}_\xi^M) = \omega^\xi$.

In the case ξ is a limit ordinal, if $(\xi_n)_{n\in\mathbb{N}}$ is the strictly increasing sequence of ordinals in the definition of $\mathcal{S}_\xi$ then for each n

$$\emptyset \neq (\mathcal{S}_\xi^M)^{(\omega^{\xi_n})} \subset \Big\{F \subset \mathbb{N} : \ \exists k > n \text{ with } k \leq F \text{ and } F \in \mathcal{S}_{\xi_k}^M\Big\}$$

which yields that $(\mathcal{S}_\xi^M)^{\omega^\xi} = \{\emptyset\}$ thus $i(\mathcal{S}_\xi^M) = \omega^\xi$. □

NOTATION 12.7. For a family $\mathcal{M}$ of finite subsets of $\mathbb{N}$ and an $n \in \mathbb{N}$ we use the notation:

$$\mathcal{M}^n = \underbrace{\mathcal{M} * \mathcal{M} * \cdots * \mathcal{M}}_{n \text{ times}}.$$

LEMMA 12.8. Let $\eta, \xi < \omega_1$ and $q < \omega$ such that $\eta \cdot q < \xi$. Then for every $N \in [\mathbb{N}]$ there exists $L \in [N]$ such that $(\mathcal{S}_\eta^M)^q[L] \subset \mathcal{S}_\xi$ and $\mathcal{S}_\eta^q[L] \subset \mathcal{S}_\xi$.

Proof. By Proposition 12.1 and Lemma 12.6 we have that

$$i((\mathcal{S}_\eta^M)^q) \leq \underbrace{i(\mathcal{S}_\eta^M) \cdot i(\mathcal{S}_\eta^M) \cdot \ldots \cdot i(\mathcal{S}_\eta^M)}_{q \text{ times}} = \underbrace{\omega^\eta \cdot \omega^\eta \cdot \ldots \cdot \omega^\eta}_{q \text{ times}} = \omega^{\eta \cdot q} < \omega^\xi$$

thus $i((\mathcal{S}_\eta^M)^q) < i(\mathcal{S}_\xi)$. Corollary 12.4 yields the existence of an $L \in [N]$ such that $(\mathcal{S}_\eta^M)^q[L] \subset \mathcal{S}_\xi$.

Since $\mathcal{S}_\eta \subset \mathcal{S}_\eta^M$ we also get that $\mathcal{S}_\eta^q[L] \subset \mathcal{S}_\xi$. □

The following proposition is proved in [**10**] (Proposition 2.15 of [**10**]).

PROPOSITION 12.9. For all ordinals $\eta < \zeta < \omega_1$ and every $M \in [\mathbb{N}]$, $\varepsilon > 0$ there exists $N \in [M]$ such that

$$\sup_{F \in \mathcal{S}_\eta} \sum_{k \in F} \zeta_1^L(k) < \varepsilon \quad \text{for every} \quad L \in [N].$$

The next proposition will help us to verify the existence of certain vectors (B.S.C.C. and S.C.C.) in the spaces we deal with in the next chapter.

PROPOSITION 12.10. Let $\eta, \zeta, \xi < \omega_1$ and $q \in \mathbb{N}$ be such that $\eta \cdot q + 1 < \zeta \leq \xi$. Then for every $M \in [\mathbb{N}]$ and $\varepsilon > 0$ there exists $N \in [M]$ such that for every $L \in [N]$ the following conditions are fulfilled:

(i) $\operatorname{supp} \zeta_1^L \in \mathcal{S}_\xi$.

(ii) $\sum_{k \in F} \zeta_1^L(k) < \varepsilon$ for all $F \in (\mathcal{S}_\eta^M)^q$.

Proof. By Lemma 12.8 there exists $M' \in [M]$ such that $(\mathcal{S}_\eta^M)^q[M'] \subset \mathcal{S}_{\eta \cdot q+1}$. Since $\eta \cdot q + 1 < \zeta$, Proposition 12.9 yields that there exists $N' \in [M']$ such that $\sup_{F \in \mathcal{S}_{\eta \cdot q+1}} \sum_{k \in F} \zeta_1^L(k) < \varepsilon$ for all $L \in [N']$. Corollary 12.4 also yields that there exists $N \in [N']$ such that $\mathcal{S}_\zeta[N] \subset \mathcal{S}_\xi$. We claim that the set N is the desired set of the statement of the proposition.

Indeed, let $L \in [N]$. Then $\operatorname{supp} \zeta_1^L \in \mathcal{S}_\zeta[L] \subset \mathcal{S}_\zeta[N] \subset \mathcal{S}_\xi$ so property (i) is fulfilled. Let also $F \in (\mathcal{S}_\eta^M)^q$. Then $F \cap L \subset L \subset N \subset N' \subset M'$ and $F \cap L \in (\mathcal{S}_\eta^M)^q$ thus $F \cap L \in \mathcal{S}_{\eta \cdot q+1}$. Therefore

$$\sum_{k \in F} \zeta_1^L(k) = \sum_{k \in F \cap L} \zeta_1^L(k) < \varepsilon$$

so property (ii) is fulfilled. □

CHAPTER 13

The space $X_G = T\left[G, (\mathcal{S}_{\xi_j}, \frac{1}{m_j})_j, D\right]$ for an $\mathcal{S}_\xi$ bounded set G

This chapter contains a presentation of the results concerning the space $X_G = T[G, (\mathcal{S}_{\xi_j}, \frac{1}{m_j})_j, D]$ for an $\mathcal{S}_\xi$ bounded set G. As the reader will see almost all the results are obtained by repeating the arguments of the corresponding results for $T[G, (\mathcal{S}_{n_j}, \frac{1}{m_j})_j, D]$. The only point where we have to use a slightly different argument concerns those proofs which involve (ε, j) S.C.C. There we proceed with use of Proposition 12.10 which requires an initial choice of a subset N of M (M is given) inside in which we restrict our constructions.

DEFINITION 13.1. A subset G of c_{00} is said to be $\mathcal{S}_\xi$ bounded for a $\xi < \omega_1$ if the following are fulfilled:

(1) G is a ground set (Definition 3.1).
(2) The ground space Y_G (Definition 3.2) does not contain isomorphically ℓ^1.
(3) For every weakly null block sequence $(y_n)_{n\in\mathbb{N}}$ in Y_G and every $\varepsilon > 0$ there exists an $L \in [\mathbb{N}]$ such that for every $g \in G$

$$\{\min\operatorname{supp} y_n : n \in L \text{ such that } |g(y_n)| > \varepsilon\} \in \mathcal{S}_\xi.$$

REMARK 13.2. Rosenthal's ℓ^1 Theorem yields that if G is $\mathcal{S}_\xi$ bounded then for every bounded block sequence $(x_n)_{n\in\mathbb{N}}$ in Y_G and every $\varepsilon > 0$ there exists a subsequence $(x_{l_n})_{n\in\mathbb{N}}$ of $(x_n)_{n\in\mathbb{N}}$ such that for the sequence $(y_n)_{n\in\mathbb{N}}$ with $y_n = x_{l_{2n-1}} - x_{l_{2n}}$ we have

$$\{t_n : n \in \mathbb{N} \ \ |g(y_n)| > \varepsilon\} \in \mathcal{S}_\xi$$

for every $g \in G$ (where $t_n = \min\operatorname{supp} y_n$ for each n).

LEMMA 13.3. If $\zeta < \xi < \omega_1$ then there exists a finite ordinal t such that $\mathcal{S}_\zeta \subset \mathcal{S}_{\xi+t}$.

Proof. Let $k \in \mathbb{N}$ be such that if $F \in \mathcal{S}_\zeta$ with $k \leq F$ then $F \in S_\xi$. Let t be a finite ordinal such that $\{2, 3, \ldots, k\} \in \mathcal{S}_t$. It follows that $\mathcal{S}_\zeta \subset \mathcal{S}_t * \mathcal{S}_\xi = \mathcal{S}_{\xi+t}$. $\square$

DEFINITION 13.4 (The space $\boldsymbol{X_G = T\left[G, (\mathcal{S}_{\xi_j}, \frac{1}{m_j})_j, D\right]}$). Let G be an $\mathcal{S}_\xi$ bounded set and let $(m_j)_{j\in\mathbb{N}}$ be the sequence of natural numbers defined by

$$m_1 = 2 \qquad m_{j+1} = {m_j}^{m_j}$$

We first recursively choose a strictly increasing sequence of ordinals $(\xi_j)_{j\in\mathbb{N}}$ in the following manner: Set $\xi_1 = \xi$ and if ξ_j has been defined select $\xi_{j+1} > (\xi_j + 1)\cdot 2^{2m_{j+1}}$ such that $\mathcal{S}_{\xi_{j+1}} \supset \mathcal{S}_{\xi_j}$. Such a choice is possible (Lemma 13.3).

The set $W = W\left[G, (\mathcal{S}_{\xi_j}, \frac{1}{m_j})_j\right]$ and the corresponding set A are defined as in Definition 3.3, replacing each n_j by ξ_j. A finite (resp. infinite) sequence $(f_i)_{i=1,\dots,k}$ (resp. $(f_i)_{i\in\mathbb{N}}$) of successive functionals in A is said to be a *special sequence* if

(i) $w(f_1) \in \{m_{2j} : j \in \mathbb{N}\}$

(ii) For each $i = 1, \dots, k-1$ (resp. $i \in \mathbb{N}$) $w(f_{i+1}) = m_{\sigma(f_1,\dots,f_i)}$.

where σ is the Gowers-Maurey type injection described in Definition 7.1.

The norming set D of the space $X_G = T\left[G, (\mathcal{S}_{\xi_j}, \frac{1}{m_j})_j, D\right]$ is the set $D = \bigcup_{n=0}^{\infty} D_n$ where the sets D_n are defined as follows:

$$C_0 = G, \qquad D_0 = \operatorname{conv}_{\mathbb{Q}}(C_0)$$

$$\begin{aligned} C_{n+1} &= C_n \cup \bigcup_{j=1}^{\infty} \Big\{ \frac{1}{m_{2j}} \sum_{i=1}^{d} f_i : \ f_i \in D_n \text{ for each } i = 1, 2, \dots, d \\ &\qquad \text{and the sequence } (f_1, \dots, f_d) \text{ is } \mathcal{S}_{\xi_{2j}} \text{ admissible} \Big\} \\ &\cup \bigcup_{j=1}^{\infty} \Big\{ \pm E(\frac{1}{m_{2j-1}} \sum_{i=1}^{d} f_i) : \ (f_1, \dots, f_d) \text{ is an } \mathcal{S}_{\xi_{2j-1}} \text{ admissible special} \\ &\qquad \text{sequence in } C_n \text{ with } w(f_1) > m_{2j-1} \text{ and } E \text{ is an interval of } \mathbb{N} \Big\} \end{aligned}$$

$$D_{n+1} = \operatorname{conv}_{\mathbb{Q}}(C_{n+1})$$

DEFINITION 13.5 (The auxiliary space $\boldsymbol{T}$). The auxiliary space T in this case is the space

$$T = T[G_0, (\mathcal{S}^M_{\xi_j+1}, \frac{1}{m_j})_j]$$

where $G_0 = \{\pm e_q^* : q \in \mathbb{N}\}$ and $\mathcal{S}^M_\beta$ denotes the β $\underline{th}$ modified Schreier family.

The norming set of the space T may alternatively described (as in Definition 3.14) as $B = \bigcup_{n=0}^{\infty} B_n$ where the sequence $(B_n)_{n\in\mathbb{N}}$ is recursively defined as follows:

$$\begin{aligned} B_0 &= \{\pm e_q : \ q \in \mathbb{N}\} \\ B_{n+1} &= \Big\{ \frac{1}{m_j}(f_1 + f_2 + \dots + f_d) : \quad f_i \in B_n \ \text{ for } i = 1, 2, \dots, d \\ &\qquad \text{and the sequence } (f_1, f_2, \dots, f_d) \text{ is } \mathcal{S}^M_{\xi_j+1} \text{ admissible, } j \in \mathbb{N} \Big\}. \end{aligned}$$

The reason we use the modified Schreier families in the definition of the auxiliary space T is the following. In the proof of Proposition 13.8 (which corresponds to Proposition 4.3) we encounter the union $F = \bigcup_{i=1}^{7} F_i$ where $F_i \in \mathcal{S}_{\xi_j}$ with $\min F_i \geq 7$. Clearly the family $\mathcal{S}^M_{\xi_j+1}$ contains the set F. This enforces us to use the modified Schreier families in the definition of the B.S.C.C and S.C.C. in this case.

DEFINITION 13.6. Let $(x_k)_{k\in\mathbb{N}}$ be a block sequence and $\varepsilon > 0$. We set $t_k = \min\operatorname{supp} x_k$ for each $k = 1, 2, \dots$ The sequence $(x_k)_{k\in\mathbb{N}}$ is said to be (G, ε, ξ) Schreier if for every $f \in G$ we have

$$\{t_k : \ k \in \mathbb{N} \text{ such that } |f(x_k)| > \varepsilon\} \in \mathcal{S}_\xi.$$

DEFINITION 13.7. Let $C \geq 1$ and $\varepsilon > 0$. A block sequence $(x_k)_{k\in\mathbb{N}}$ in X_G is said to be a (C, ε, ξ) *rapidly increasing sequence* ((C, ε, ξ) R.I.S.) if the following conditions hold

(1) The sequence $(x_k)_{k\in\mathbb{N}}$ is (G, ε, ξ) Schreier with $\|x_k\| \leq C$ for all k, while $\min\operatorname{supp} x_1 \geq 7$.
(2) There exists a strictly increasing sequence $(j_k)_{k\in\mathbb{N}}$ such that:
  (a) $(\max\operatorname{supp} x_k)\frac{1}{\varepsilon} \leq m_{j_{k+1}}$ for each k.
  (b) For each $k = 1, 2, \ldots$ and $h \in A \cap D$ with $w(h) < m_{j_k}$ we have $|h(x_k)| \leq \dfrac{C}{w(f)}$.

Proceeding as in chapter 4 we may prove the following proposition, the statement of which, as well as its proof, is analogous to Proposition 4.3.

PROPOSITION 13.8 (Basic inequality). Let $(x_k)_{k\in\mathbb{N}}$ be a (C, ε, ξ) R.I.S. in X_G. Then for every $f \in A \cap D$ we can find g_1, such that either $g_1 = h_1$ or $g_1 = e_t^* + h_1$ with $t \notin \operatorname{supp} h_1$ where $h_1 \in \operatorname{conv}_{\mathbb{Q}}\{h \in B : \; w(h) = w(f)\}$ and $g_2 \in c_{00}$ with $\|g_2\|_\infty \leq \varepsilon$ such that

$$|f(\sum \lambda_k x_k)| \leq C(2g_1 + g_2)(\sum |\lambda_k| e_{t_k}) \tag{38}$$

where $t_k = \min\operatorname{supp} x_k$ for each $k = 1, 2, \ldots$.

If we additionally assume that for some $j_0 \in \mathbb{N}$ the next condition is satisfied: For every $h \in A \cap D$ with $w(h) = m_{j_0}$ and every interval E of the natural numbers holds

$$|h(\sum_{k\in E} \lambda_k x_k)| \leq C(2\max_{k\in E} |\lambda_k| + \varepsilon \sum_{k\in E} |\lambda_k|) \tag{39}$$

then h_1 may be selected satisfying the following property: $h_1 = \sum r_i \tilde{h}_i$ with $r_i \in \mathbb{Q}^+$, $\sum r_i = 1$ and for each i the functional $\tilde{h}_i$ belongs to B and has a tree $T_{\tilde{h}_i} = (f_a^i)_{a\in\mathcal{C}_i}$ with $w(f_a^i) \neq m_{j_0}$ for all $a \in \mathcal{C}_i$.

For the rest of this chapter we set $q_j = 2\log_2(m_j)$ for each j.

DEFINITION 13.9 (basic special convex combinations and special convex combinations). Let $\varepsilon > 0$ and $j \in \mathbb{N}$, $j > 1$. A convex combination $\sum\limits_{k\in F} a_k e_k$ of the basis $(e_k)_{k\in\mathbb{N}}$ is said to be an (ε, j) *basic special convex combination* ((ε, j) B.S.C.C.) if

(1) $F \in \mathcal{S}_{\xi_j}$.
(2) For every $P \in (\mathcal{S}^M_{\xi_{j-1}+1})^{q_j}$ we have that $\sum\limits_{k\in P} a_k < \varepsilon$.
(3) The sequence $(a_k)_{k\in F}$ is in decreasing order.

If $(x_k)_{k\in\mathbb{N}}$ is a block sequence we say that $\sum\limits_{k\in F} a_k x_k$ is an (ε, j) *special convex combination* (S.C.C.) if the corresponding convex combination of the basis $\sum\limits_{k\in F} a_k e_{t_k}$, where $t_k = \min\operatorname{supp} x_k$ for each k, is an (ε, j) B.S.C.C. If moreover the sequence $(x_k)_{k\in\mathbb{N}}$ is such that $\|x_k\| \leq 1$ and $\|\sum\limits_{k\in F} a_k x_k\| \geq \frac{1}{2}$ then $\sum\limits_{k\in F} a_k x_k$ is said to be a *seminormalized* (ε, j) S.C.C.

LEMMA 13.10 (Existence of B.S.C.C). Let $\varepsilon > 0$, $j \in \mathbb{N}$ with $j > 1$ and let ζ be an ordinal such that $(\xi_{j-1}+1)\cdot q_j + 1 < \zeta \leq \xi_j$. Then for every $M \in [\mathbb{N}]$ there exists $N \in [M]$ such that for every $L \in [N]$ the vector ζ_1^L is an (ε, j) B.S.C.C.

Proof. This follows readily from Proposition 12.10 and the definition of the B.S.C.C. □

Proceeding as in chapter 3 we may prove the following proposition for estimates on B.S.C.C. in the space T. The proof of this proposition is similar to the proof of Proposition 3.19 and relies on a lemma analogous to Lemma 3.16. Note that since the sequence $(\xi_i)_{i\in\mathbb{N}}$ has been selected to satisfy $\mathcal{S}_{\xi_i} \subset \mathcal{S}_{\xi_{i+1}}$ we also have that $\mathcal{S}^M_{\xi_i+1} \subset \mathcal{S}^M_{\xi_{i+1}+1}$.

PROPOSITION 13.11 (Estimates on the basis of T). Let $\varepsilon > 0$, $j \in \mathbb{N}$ with $j > 1$ such that $\varepsilon \leq \frac{1}{m_j}$ and let $x = \sum_{k\in F} a_k e_k$ be an (ε, j) B.S.C.C. Then:

(i) For every $f \in B$

$$|f(x)| \leq \begin{cases} \varepsilon & \text{if } f = \pm e_q^* \\ \dfrac{1}{m_i} & \text{if } w(f) = m_i \ , i \geq j \\ \dfrac{2}{m_i m_j} & \text{if } w(f) = m_i \ , i < j \end{cases}$$

and thus $|f(x)| \leq \frac{1}{m_j}$.

(ii) If the functional $f \in B$ has a tree $T_f = (f_a)_{a\in\mathcal{A}}$ with $w(f_a) \neq m_j$ for all $a \in \mathcal{A}$ and $\varepsilon \leq \frac{1}{m_j^2}$ then $|f(x)| \leq \frac{1}{m_j^2}$.

The following corollary is an immediate consequence of Lemma 13.10.

COROLLARY 13.12 (Existence of S.C.C.). For every block sequence $(x_k)_{k\in\mathbb{N}}$, $\varepsilon > 0$, $j \in \mathbb{N}$ with $j > 1$ and every ordinal ζ such that $(\xi_{j-1}+1)\cdot q_j + 1 < \zeta \leq \xi_j$ there exists an $N \in [\mathbb{N}]$ such that for every $L \in [N]$ the vector $a(\zeta, (x_k)_{k\in L})$ is an (ε, j) S.C.C.

The following proposition ensures the existence of seminormalized special convex combinations in every block subspace of X_G.

PROPOSITION 13.13. Let $(x_k)_{k\in\mathbb{N}}$ be a normalized block sequence in X_G, $\varepsilon > 0$ and let $j \in \mathbb{N}$ be an even natural number (in order the norming set D to be j closed i.e. if $(f_1, \dots, f_d)$ is an $\mathcal{S}_{\xi_j}$ admissible sequence of D then the functional $\frac{1}{m_j}(f_1 + \cdots + f_d)$ belongs to D). Then there exists a block sequence $(y_k)_{k\in\mathbb{N}}$ of $(x_k)_{k\in\mathbb{N}}$ with $\|y_k\| \leq 1$ for all k and an (ε, j) S.C.C. $\sum_{k\in F} a_k y_k$ such that

(1) $\|\sum_{k\in F} a_k y_k\| \geq \frac{1}{2}$.

(2) The family $\{y_k : k \in F\}$ is $\mathcal{S}_{(\xi_{j-1}+1)(2q_j+2)}$ admissible.

Proof. Set $t_k = \min \operatorname{supp} x_k$ for $k = 1, 2, \dots$ and $T = \{t_k : k \in \mathbb{N}\}$. We recall that $q_j = 2\log_2(m_j)$ and observe that $(\xi_{j-1}+1)(2q_j+2)q_j < \xi_j$. Lemma 12.8 yields the existence of a $T' \in [T]$ such that $\mathcal{S}^{q_j}_{(\xi_{j-1}+1)(2q_j+2)}[T'] \subset \mathcal{S}_{\xi_j}$. We may assume, passing to a subsequence of $(x_k)_{k\in\mathbb{N}}$, that $\mathcal{S}^{q_j}_{(\xi_{j-1}+1)(2q_j+2)}[T] \subset \mathcal{S}_{\xi_j}$.

The proof continues as the proof of Proposition 5.4. We assume the contrary and construct block sequences $(y^l_k)_{k\in\mathbb{N}}$ for $0 \leq l \leq q_j$ satisfying

(i) For $1 \leq l \leq q_j$ $(y^l_k)_{k\in\mathbb{N}}$ is a block sequence of $(y^{l-1}_k)_{k\in\mathbb{N}}$ and each y^l_k is a convex combination of the sequence $(x_k)_{k\in\mathbb{N}}$.

(ii) For all l and k the family $\{\operatorname{supp} x_i : \operatorname{supp} x_i \subset \operatorname{supp} y^l_k\}$ is $\mathcal{S}^l_{(\xi_{j-1}+1)(2q_j+2)}$ admissible.

(iii) $\dfrac{1}{m_j} \leq \|y^l_k\| < \dfrac{1}{2^l}$ for $1 \leq l \leq q_j$ and all k.

We then arrive to a contradiction since (iii) yields that $2^{q_j} < m_j$ i.e. $m_j < m_j$. ☐

The following proposition is a consequence of Proposition 13.8 and Lemma 13.11. The statement of Proposition 13.14 as well as its proof are identical to those of the corresponding Proposition 6.2.

PROPOSITION 13.14. Let $C \geq 1$, $j \in \mathbb{N}$ with $j > 1$ and let δ, ε be positive real numbers such that $\max\{\delta,\varepsilon\} \leq \dfrac{1}{m_j}$. Also let $(x_k)_{k\in\mathbb{N}}$ be a (C,ε,ξ) rapidly increasing sequence of X_G. Then:

(1a) If $\sum\limits_{k\in F} a_k x_k$ is a (δ, j) S.C.C. of the sequence $(x_k)_{k\in\mathbb{N}}$ then

$$\|\sum_{k\in F} a_k x_k\| \leq \frac{5C}{m_j}.$$

(1b) If moreover $\max\{\delta,\varepsilon\} \leq \dfrac{1}{m_j^2}$ then for every $f \in A \cap D$ with $w(f) < m_j$ we have that

$$|f(\sum_{k\in F} a_k x_k)| \leq \frac{7C}{w(f) m_j}.$$

(2) If $\max\{\delta,\varepsilon\} \leq \dfrac{1}{m_j^2}$, the sequence $(a_k)_{k\in F}$ is such that $\sum\limits_{k\in F} |a_k| x_k$ is a (δ, j) S.C.C. and for every $h \in D$ with $w(h) = m_j$ and every interval E of the natural numbers we have that

$$|h(\sum_{k\in E} a_k x_k)| \leq C(2\max_{k\in E\cap F} |a_k| + \varepsilon \sum_{k\in E\cap F} |a_k|)$$

then

$$\|\sum_{k\in F} a_k x_k\| \leq \frac{5C}{m_j^2}$$

We also may prove the following analogue to Proposition 5.3, for making estimates on special convex combinations.

PROPOSITION 13.15. Let $(x_k)_{k\in\mathbb{N}}$ be a block sequence in X_G with $\|x_k\| \leq C$ for each k, $\varepsilon > 0$ and let $j \in \mathbb{N}$ with $j > 1$ such that $\varepsilon \leq \dfrac{1}{m_j}$, and $\sum\limits_{k\in F} a_k x_k$ an

(ε, j) S.C.C.
Then for every $i < j$ and $f \in A \cap D$ with $w(f) = m_i$ we have that

$$|f(\sum_{k \in F} a_k x_k)| \leq \frac{2C}{m_i}.$$

The following lemma, the statement of which as well as its proof are similar to Lemma 6.5, verifies the existence of certain rapidly increasing sequences.

LEMMA 13.16. Let G be an $\mathcal{S}_\xi$ bounded set, $(x_k)_{k\in\mathbb{N}}$ a block sequence in $X_G = T[G, (\mathcal{S}_{\xi_j}, \frac{1}{m_j})_j, D]$ and $\varepsilon > 0$. Suppose also that there exists a strictly increasing sequence $(i_k)_{k\in\mathbb{N}}$ of the natural numbers such that each x_k is a seminormalized $(\frac{1}{m_{i_k}}, i_k)$ S.C.C. Then there exists a subsequence $(x_{l_k})_{k\in\mathbb{N}}$ of $(x_k)_{k\in\mathbb{N}}$ such that the sequence $(y_k)_{k\in\mathbb{N}}$ of successive differences of this sequence (i.e. $y_k = x_{l_{2k-1}} - x_{l_{2k}}$) is a $(4, \varepsilon, \xi)$ rapidly increasing sequence.

DEFINITION 13.17. A double sequence $(x_k, x_k^*)_{k\in\mathbb{N}}$ is said to be a $(\theta, 2j+1)$ *dependent couple* (for $j \in \mathbb{N}$ and $0 \leq \theta \leq 1$) if there exists a sequence $(2j_k)_{k\in\mathbb{N}}$ of even integers such that the following conditions are fulfilled:
(i) $(x_k)_{k\in\mathbb{N}}$ is a block sequence in X_G, $x_k = \theta_k m_{2j_k} \sum_{i\in F_k} a_i^k x_i^k$ where $\sum_{i\in F_k} a_i^k x_i^k = a(\xi_{2j_k}, (x_i^k)_i)$ is a $(\frac{1}{4m_{2j_k}^2}, 2j_k)$ S.C.C. of $(x_i^k)_i$, the sequence $(x_i^k)_i$ is a $(4, \frac{1}{m_{2j_k}^2})$ R.I.S. and $\theta_k \leq 1$.
(ii) $(x_k^*)_{k\in\mathbb{N}}$ is a special sequence with $w(x_k^*) = m_{2j_k}$ and $2j_1 \geq 2j+4$.
(iii) $x_k^*(x_k) = \theta$ and $\operatorname{ran} x_k^* = \operatorname{ran} x_k$ for all k.

The statement of the next proposition, as well as its proof, is analogous to Proposition 7.5. The only difference is that in Proposition 7.5 the assumption $\min\operatorname{supp} x_1 > 3m_{2j+1}^2$ guarantees that $a(n_{2j+1}, (x_k)_{k\in\mathbb{N}})$ is a $(\frac{1}{m_{2j+1}^2}, 2j+1)$ S.C.C. while in Proposition 13.18 we have to assume in the statement that $a(\xi_{2j+1}, (x_k)_{k\in\mathbb{N}})$ is a $(\frac{1}{m_{2j+1}^2}, 2j+1)$ S.C.C.

PROPOSITION 13.18. Let $(x_k, x_k^*)_{k\in\mathbb{N}}$ be a $(\theta, 2j+1)$ dependent couple. If the convex combination $a(\xi_{2j+1}, (x_k)_{k\in\mathbb{N}}) = \sum_{k=1}^{n} a_k x_k$ is a $(\frac{1}{m_{2j+1}^2}, 2j+1)$ S.C.C. of the sequence $(x_k)_{k\in\mathbb{N}}$ then

$$\|\sum_{k=1}^{n} (-1)^k a_k x_k\| \leq \frac{140}{m_{2j+1}^2}.$$

THEOREM 13.19. *If the ground set G is $\mathcal{S}_\xi$ bounded then the space $X_G = T\left[G, (\mathcal{S}_{\xi_j}, \frac{1}{m_j})_j, D\right]$ is H.I.*

Proof. It is enough, for every pair Y, Z of subspaces of X_G and every $\delta > 0$, to find $y \in Y$ and $z \in Z$ with $\|y - z\| < \delta\|y + z\|$. By standard perturbation arguments we may assume that $Y = \overline{\operatorname{span}}\{y_m : m \in \mathbb{N}\}$ and $Z = \overline{\operatorname{span}}\{z_m : m \in \mathbb{N}\}$ where $(y_m)_{m\in\mathbb{N}}$ and $(z_m)_{m\in\mathbb{N}}$ are block sequences in X_G with $\operatorname{ran} y_m = \operatorname{ran} z_m$.

Let Y, Z, δ be as before and choose $j \in \mathbb{N}$ with $m_{2j+1} > \frac{280}{\delta}$ and set $\varepsilon = \frac{1}{m_{2j+1}^2}$. We set $M = \{t_m : m \in \mathbb{N}\}$ where $t_m = \min \operatorname{supp} y_m = \min \operatorname{supp} z_m$ for $m = 1, 2, \ldots$ By Lemma 13.10 there exists $N \in [M]$ such that for all $L \in [N]$ the vector $(\xi_{2j+1})_1^L$ is an $(\varepsilon, 2j+1)$ B.S.C.C. Passing to subspaces of Y,Z, spanned by subsequences of $(y_m)_{m\in\mathbb{N}}$ and $(z_m)_{m\in\mathbb{N}}$ respectively, we may assume that the vector $(\xi_{2j+1})_1^L$ is an $(\varepsilon, 2j+1)$ B.S.C.C. for every $L \in [M]$.

The rest of the proof continues as the corresponding proof of Theorem 7.3. Thus we find a dependent couple $(x_k, x_k^*)_{k\in\mathbb{N}}$ with $x_k \in Y$ if k is odd and $x_k \in Z$ if k is even and then, by our previous choice of the sequences $(y_m)_{m\in\mathbb{N}}$ and $(z_m)_{m\in\mathbb{N}}$ the vector $a(\xi_{2j+1}, (x_k)_{k\in\mathbb{N}}) = \sum_{k=1}^{n} a_k x_k$ is a $(\varepsilon, 2j+1)$ S.C.C. Therefore, as in the proof of Theorem 7.5, the vectors

$$y = \sum_{k\le n,\ k \text{ odd}} a_k x_k \quad\text{and}\quad z = \sum_{k\le n,\ k \text{ even}} a_k x_k.$$

satisfy $y \in Y$, $z \in Z$ and $\|y - z\| < \delta\|y + z\|$. □

The next proposition, is the analogue of Proposition 13.13 in terms of the predual space $(X_G)_*$.

PROPOSITION 13.20. Let $(f_k)_{k\in\mathbb{N}}$ be a block sequence of the basis $(e_k^*)_{k\in\mathbb{N}}$ of $(X_G)_*$ such that $f_k \in D$ and $\|f_k\| \ge \frac{1}{2}$ for all k. Also let $j \in \mathbb{N}$ be an even integer. Then there exists a finite block sequence $(h_k)_{k\in F}$ of $(f_k)_{k\in\mathbb{N}}$ such that

(1) $\|h_k\| > \frac{1}{2}$ for all k and $\|\sum_{k\in F} h_k\| \le 1$.
(2) The set $\{\min \operatorname{supp} h_k : k \in F\}$ is a maximal $\mathcal{S}_{(\xi_{j-1}+1)(2q_j+2)}$ set.
(3) There exists a $c < 1$ such that each h_k is of the form $h_k = c \sum_{i\in F_k} f_i$.

Proof. We may assume, passing to a subsequence of $(f_k)_{k\in\mathbb{N}}$ (as in the proof of Proposition 13.13), that $\mathcal{S}^{q_j}_{(\xi_{j-1}+1)(2q_j+2)}[T] \subset \mathcal{S}_{\xi_j}$, whwre $T = \{\min \operatorname{supp} f_k : k \in \mathbb{N}\}$. The rest of the proof is exactly the same to that of Proposition 5.5. □

LEMMA 13.21. Let Z be a block subspace of the basis $(e_k^*)_{k\in\mathbb{N}}$ of $(X_G)_*$, $\delta > 0$ and $j \in \mathbb{N}$ with $j > 1$. Then there exists a vector $x \in X_G$ and a functional $h \in D$ such that

(i) $\operatorname{dist}(h, Z) < \delta$.
(ii) The vector x is a $(\frac{1}{m_{2j}}, 2j)$ S.C.C. of a normalized block sequence.
(iii) $h(x) > \frac{1}{3}$ and $\operatorname{ran} h = \operatorname{ran} x$.

Proof. We set $\eta = (\xi_{2j-1} + 1)(2q_{2j} + 2)$. By Lemma 13.10 we can choose a normalized block sequence $(z_k^*)_{k\in\mathbb{N}}$ in Z such that the set $N = \{\min \operatorname{supp} z_k^* : k \in \mathbb{N}\}$ satisfies the property that for every $L \in [N]$ the vector η_1^L is a $(\frac{1}{m_{2j}}, 2j)$ B.S.C.C. The proof is continued in the same manner as the proof of the corresponding Lemma 8.2. □

THEOREM 13.22. *If the ground set G is $\mathcal{S}_\xi$ bounded then the predual $(X_G)_* = \overline{\operatorname{span}}\{e_n^* : n \in \mathbb{N}\}$ of the space $X_G = T\left[G, (\mathcal{S}_{\xi_j}, \frac{1}{m_j})_j, D\right]$ is H.I.*

Proof. Let Y, Z be a pair of subspaces of $(X_G)_*$ and $\delta > 0$. We have to find $h_Y \in Y$ and $h_Z \in Z$ with $\|h_Y - h_Z\| < \delta \|h_Y + h_Z\|$. We select $j \in \mathbb{N}$ with $m_{2j+1} > 420(\frac{2}{\delta} + 1)$ and set $\varepsilon = \frac{1}{m_{2j+1}^2}$.

By standard perturbation arguments we may assume that $Y = \overline{\text{span}}\{y_m^* : m \in \mathbb{N}\}$, $Z = \overline{\text{span}}\{z_m^* : m \in \mathbb{N}\}$ where $(y_m^*)_{m\in\mathbb{N}}$ and $(z_m^*)_{m\in\mathbb{N}}$ are block sequences in $(X_G)_*$ with $\text{ran}\, y_m^* = \text{ran}\, z_m^*$. Using the same argument as in the proof of Theorem 13.19 (by Lemma 13.10) we may also assume that the set $M = \{\min\text{supp}\, y_m^* : m \in \mathbb{N}\}$ satisfies the property that for all $L \in [M]$ the vector $(\xi_{2j+1})_1^L$ is an $(\varepsilon, 2j+1)$ B.S.C.C.

The rest of the proof is similar to the proof of the corresponding Theorem 8.3. □

LEMMA 13.23. Suppose that the ground set G is $\mathcal{S}_\xi$ bounded. Let $(y_k)_{k\in\mathbb{N}}$ be a bounded block sequence in X_G and Y a closed subspace of X_G containing the sequence $(y_k)_{k\in\mathbb{N}}$. Let also $T : Y \longrightarrow X_G$ be a bounded linear operator satisfying the following conditions:

(1) There exists a strictly increasing sequence $(i_k)_{k\in\mathbb{N}}$ of integers such that each y_k is a seminormalized $(\frac{1}{m_{2i_k}}, 2i_k)$ S.C.C.
(2) $(Ty_k)_{k\in\mathbb{N}}$ is a block sequence in X_G.

Then:

$$\text{dist}(Ty_n, \mathbb{R}y_n) \overset{n\to\infty}{\longrightarrow} 0.$$

Proof. We assume the contrary. As in the proof of the corresponding Lemma 9.1 we may assume that there exist $\delta > 0$ and a sequence $(y_n^*)_{n\in\mathbb{N}}$ in D such that $\text{ran}\, y_n = \text{ran}\, Ty_n = \text{ran}\, y_n^*$, $y_n^*(Ty_n) > \delta$ and $y_n^*(y_n) = 0$. We then choose $j \in \mathbb{N}$ with $m_{2j+1} > \frac{140 \cdot \|T\|}{\delta}$. Passing to a subsequence we may assume (by Lemma 13.10) that the set $M = \{\min\text{supp}\, y_n : n \in \mathbb{N}\}$ satisfies the property that for all $L \in [M]$ the vector $(\xi_{2j+1})_1^L$ is a $(\frac{1}{m_{2j+1}^2}, 2j+1)$ B.S.C.C.

The proof continues in the same way as the proof of Lemma 9.1. □

The proof of the next Theorem is identical to that of Theorem 9.2.

THEOREM 13.24. *Let G be a ground $\mathcal{S}_\xi$ bounded set. Then every bounded linear operator in $\mathcal{L}(X_G)$ takes the form $T = \lambda I + S$ where $\lambda \in \mathbb{R}$ and S is a strictly singular operator.*

The next Lemma corresponds to Lemma 9.3.

LEMMA 13.25. Suppose that the ground set G is $\mathcal{S}_\xi$ bounded. Let $(y_k)_{k\in\mathbb{N}}$ be a bounded block sequence in X_G and Y a closed subspace of X_G containing the sequence $(y_k)_{k\in\mathbb{N}}$. Let also $T : Y \longrightarrow X_G$ be a bounded linear operator satisfying the following conditions.

(1) There exists a strictly increasing sequence $(i_k)_{k\in\mathbb{N}}$ of integers such that each y_k is a $(\frac{1}{m_{2i_k}}, 2i_k)$ S.C.C. of a normalized sequence and $\|y_k\| > \delta > 0$ for all k.
(2) The sequence $(w_k)_{k\in\mathbb{N}}$ defined by $w_k = y_{2k-1} - y_{2k}$ satisfies the property that $(Tw_k)_{k\in\mathbb{N}}$ is a block sequence in X_G.

Then:

$$\mathrm{dist}(Tw_n, \mathbb{R}w_n) \overset{n\to\infty}{\longrightarrow} 0.$$

REMARK 13.26. It follows from the preceding lemma that every block subspace Y of X_G contains a seminormalized weakly null sequence $(y_n)_{n\in\mathbb{N}}$ satisfying the following property. For every bounded linear operator $T \in \mathcal{L}(X_G)$ we have that $\mathrm{dist}(Ty_n, \mathbb{R}y_n) \overset{n\to\infty}{\longrightarrow} 0$. If moreover the operator T is strictly singular then $\|Ty_n\| \overset{n\to\infty}{\longrightarrow} 0$.

The proof of the next theorem relies on Lemma 13.25 and is exactly the same as the corresponding proof of Theorem 9.4.

THEOREM 13.27. *Let G be a ground set which is $\mathcal{S}_\xi$ bounded. Then every bounded linear operator in $\mathcal{L}(X_G)$ is of the form $T = \lambda I + S$ where $\lambda \in \mathbb{R}$ and S is a weakly compact operator.*

PROPOSITION 13.28. Suppose that the ground set G is $\mathcal{S}_\xi$ bounded and let $(z_n)_{n\in\mathbb{N}}$ be a normalized weakly null block sequence in $X_G = T[G, (\mathcal{S}_{n_j}, \frac{1}{m_j})_j, D]$. Then there exists $L \in [\mathbb{N}]$ such that the space $Z = \overline{\mathrm{span}}\{z_n : n \in L\}$ is reflexive.

Proof. The proof is obtained following exactly the same procedure as in the proof of Theorem 9.7, replacing only $\mathcal{S}_2$ by $\mathcal{S}_\xi$. □

Proposition 13.28 immediately yields the following:

THEOREM 13.29. *If the ground set G is $\mathcal{S}_\xi$ bounded then the space $X_G = T[G, (\mathcal{S}_{n_j}, \frac{1}{m_j})_j, D]$ is somewhat reflexive.*

THEOREM 13.30. *For every $\xi < \omega_1$ there exists a strictly increasing sequence $(\xi_j)_{j\in\mathbb{N}}$ of countable ordinals with $\xi_1 = \xi$, such that for every $\mathcal{S}_\xi$ bounded set G the space $X_G = T[G, (\mathcal{S}_{\xi_j}, \frac{1}{m_j})_j, D]$ has the following properties*

(i) *The space X_G is asymptotic ℓ^1 and $(e_n)_{n\in\mathbb{N}}$ is a bimonotone boundedly complete Schauder basis.*
(ii) *The space X_G is Hereditarily Indecomposable and somewhat reflexive.*
(iii) *The space $(X_G)_* = \overline{\mathrm{span}}\{e_n^* : n \in \mathbb{N}\}$ is Hereditarily Indecomposable.*
(iv) *The space of bounded linear operators of X_G has the form*

$$\mathcal{L}(X_G) = \{\lambda I + S : \ \lambda \in \mathbb{R}, \ S \textit{ is strictly singular and weakly compact operator }\}.$$

(v) *The space X_G^* is the norm closed linear span of the closure of G in the w^* topology of X_G^* i.e. $X_G^* = \overline{\mathrm{span}}(\overline{G}^{w^*})$.*

Proof. Assertion (i) is trivial, assertion (ii) has been proved in Theorems 13.19 and 13.29, assertion (iii) has been proved in Theorem 13.22 while assertion (iv) has been proved in Theorems 13.24 and 13.27. Finally the proof of (v) is exactly the same as the corresponding proof of Proposition 6.10. □

DEFINITION 13.31. A bounded sequence $(x_n)_{n\in\mathbb{N}}$ in a Banach space Y is an ℓ_ξ^1 *spreading model* (where $\xi < \omega_1$) if there exists a constant $C > 0$ such that $\|\sum_{i\in F} a_i x_i\| \geq C \sum_{i\in F} |a_i|$ for every $F \in \mathcal{S}_\xi$ and all choices of scalars $(a_i)_{i\in F}$.

The following Lemma follows from a result of Bourgain [**17**].

LEMMA 13.32. If a separable Banach space Y has for every $\xi < \omega_1$ an ℓ^1_ξ spreading model then Y contains isomorphically ℓ^1.

DEFINITION 13.33. A normalized sequence $(x_n)_{n\in\mathbb{N}}$ in a Banach space X is said to be *convexly unconditional* if for every $\delta > 0$ there exists $C_\delta > 0$ such that for every absolutely convex combination $x = \sum_{n=1}^{\infty} a_n x_n$, if $\|x\| \geq \delta$ then $\|\sum_{n=1}^{\infty} \varepsilon_n a_n x_n\| \geq C_\delta$ for every choice of signs $(\varepsilon_n)_{n\in\mathbb{N}}$.

One of the main results of [**13**] is that every normalized weakly null sequence in a Banach space X has a convexly unconditional subsequence.

LEMMA 13.34. Let Y be a Banach space and let $(y_n)_{n\in\mathbb{N}}$ be a convexly unconditional sequence in Y no subsequence of which is an ℓ^1_ξ spreading model. Then for every $\delta > 0$ there exists a $M \in [\mathbb{N}]$ such that for every $x^* \in B_{Y^*}$ we have that

$$\{n \in M : |x^*(y_n)| > \delta\} \in \mathcal{S}_\xi$$

Proof. Let $\delta > 0$ and set

$$\mathcal{F}_\delta = \{N \subset \mathbb{N} : \text{ there exists } x^* \in B_{Y^*} \text{ such that } |x^*(y_n)| > \delta \text{ for every } n \in N\}$$

For $L \in [\mathbb{N}]$ denote by L^ξ the maximal initial segment of L contained in $\mathcal{S}_\xi$. Set

$$\mathcal{A}_1 = \{L \in [\mathbb{N}] : L^\xi \in \mathcal{F}_\delta\}$$
$$\mathcal{A}_2 = [\mathbb{N}] \setminus \mathcal{A}_1$$

The set $\mathcal{A}_1$ is closed therefore by the infinite Ramsey Theorem we obtain a $M \in [\mathbb{N}]$ such that either $[M] \subset \mathcal{A}_1$ or $[M] \subset \mathcal{A}_2$. We observe that if the second holds then the set M satisfies the conclusion of the lemma so it is enough to exclude the possibility to have $[M] \subset \mathcal{A}_1$.

Let C_δ be the constant corresponding to δ by the convex unconditionality of the sequence $(y_n)_{n\in\mathbb{N}}$. Suppose that $[M] \subset \mathcal{A}_1$ for a set $M = \{m_i : i \in \mathbb{N}\}$. We show that the sequence $(y_{m_i})_{i\in\mathbb{N}}$ is an ℓ^1_ξ spreading model with constant C_δ.

Let $F \in \mathcal{S}_\xi$. Then $\{m_i : i \in F\} \in \mathcal{S}_\xi$ and our assumption that $[M] \subset \mathcal{A}_1$ yields that $\{m_i : i \in F\} \in \mathcal{F}_\delta$. Select an $x^* \in B_{Y^*}$ such that $|x^*(y_{m_i})| > \delta$ for all $i \in F$.

Let $(c_i)_{i\in F}$ be scalars with $\sum_{i\in F} |c_i| = 1$. We choose signs $(\varepsilon_i)_{i\in F}$ such that $x^*(\varepsilon_i c_i y_{m_i}) > 0$ for each $i \in F$. Then

$$x^*(\sum_{i\in F} \varepsilon_i c_i y_{m_i}) = \sum_{i\in F} |c_i||x^*(y_{m_i})| \geq \delta$$

It follows that $\|\sum_{i\in F} \varepsilon_i c_i y_{m_i}\| \geq \delta$ therefore $\|\sum_{i\in F} c_i y_{m_i}\| \geq C_\delta$.

We conclude that for every choice of scalars $(a_i)_{i\in F}$ we have

$$\|\sum_{i\in F} a_i y_{m_i}\| \geq C_\delta \sum_{i\in F} |a_i|$$

therefore the sequence $(y_{m_i})_{i\in\mathbb{N}}$ is an ℓ^1_ξ spreading model (with constant C_δ), a contradiction. □

PROPOSITION 13.35. Let Y be a Banach space with a Schauder basis $(e_n)_{n\in\mathbb{N}}$ such that Y does not contain any isomorphic copy of ℓ^1. Then there exists $\xi < \omega_1$ such that for every weakly null block sequence $(y_n)_{n\in\mathbb{N}}$ in Y and every $\varepsilon > 0$ there exists an $L \in [\mathbb{N}]$ such that for every $x^* \in B_{Y^*}$

$$\{\min \operatorname{supp} y_n : n \in L \text{ and } |x^*(y_n)| > \varepsilon\} \in \mathcal{S}_\xi$$

Proof. Since ℓ^1 does not embed in Y Lemma 13.32 yields a $\xi < \omega_1$ such that Y has no ℓ^1_ξ spreading model. We claim that ξ is the desired ordinal of the statement of the proposition.

Let $(y_n)_{n\in\mathbb{N}}$ be a weakly null block sequence in Y and $\varepsilon > 0$. Set $C = \sup_n \|y_n\|$. We clearly may assume that $\liminf \|y_n\| > 0$. The normalized sequence $(z_n)_{n\in\mathbb{N}}$ with $z_n = \dfrac{y_n}{\|y_n\|}$ is thus weakly null so we may assume, by passing to a subsequence, that it is convexly unconditional.

Applying Lemma 13.34 for the sequence $(z_n)_{n\in\mathbb{N}}$ and $\delta = \dfrac{\varepsilon}{C}$ we obtain an $L \in [\mathbb{N}]$ such that $\{n \in L : |x^*(z_n)| > \dfrac{\varepsilon}{C}\} \in \mathcal{S}_\xi$ for every $x^* \in \mathrm{B}_{Y^*}$. It follows that for every $x^* \in \mathrm{B}_{Y^*}$ we have $\{n \in L : |x^*(y_n)| > \varepsilon\} \in \mathcal{S}_\xi$ and the conclusion follows, since the family $\mathcal{S}_\xi$ is spreading. □

As an immediate consequence we have the following:

COROLLARY 13.36. If G is a ground (Definition 3.1) subset of c_{00} is such that the ground space Y_G (we recall that Y_G is the completion of $(c_{00}(\mathbb{N}), \| \|_G))$ contains no isomorphic copy of ℓ^1, then G is $\mathcal{S}_\xi$ bounded for some ordinal $\xi < \omega_1$.

DEFINITION 13.37. Let Y, Z be two Banach spaces with Schauder bases $(y_n)_{n\in\mathbb{N}}$ $(z_n)_{n\in\mathbb{N}}$ respectively. We say that the basis $(z_n)_{n\in\mathbb{N}}$ dominates the basis $(y_n)_{n\in\mathbb{N}}$ if there exists a constant $C > 0$ such that

$$\|\sum_{i=1}^n a_i y_i\| \leq C \|\sum_{i=1}^n a_i z_i\|$$

for every choice of scalars $(a_i)_{i=1}^n$, $n \in \mathbb{N}$ (or equivalently there exists a bounded linear operator $Q : Z \longrightarrow Y$ such that $Q(z_n) = y_n$ for all n).

THEOREM 13.38. *Let Y be a Banach space and $(y_n)_{n\in\mathbb{N}}$ a normalized bimonotone Schauder basis of Y. If ℓ^1 does not embed into Y then there exists a H.I. Banach space X with a Schauder basis $(e_n)_{n\in\mathbb{N}}$ such that the following are fulfilled:*

(i) *The basis $(e_n)_{n\in\mathbb{N}}$ of X dominates the basis $(y_n)_{n\in\mathbb{N}}$ of Y.*
(ii) *X has a predual which is also H.I.*
(iii) *Every bounded linear operator R of X can be written in the form $R = \lambda I + W$ where $\lambda \in \mathbb{R}$ and W is a strictly singular and weakly compact operator.*
(iv) *X is somewhat reflexive.*

Moreover

(a) *If Y has separable dual then X has also separable dual.*
(b) *If Y is reflexive then X is also reflexive.*

Proof. We set

$$G_0 = \Big\{\sum_{i=1}^n a_i y_i^* : \ a_i \in \mathbb{Q}, \ \|\sum_{i=1}^n a_i y_i^*\| \leq 1, \ n \in \mathbb{N}\Big\}$$

$$G = \Big\{\sum_{i=1}^n a_i e_i^* \in c_{00} : \ \sum_{i=1}^n a_i y_i^* \in G_0\Big\}$$

and observe that the set G_0 is a norming set for Y and G is a ground (Definition 3.1) subset of c_{00}. It follows from Corollary 13.36 that there exists a countable ordinal ξ such that G is $\mathcal{S}_\xi$ bounded.

The desired space X is $X_G = T\left[G, (\mathcal{S}_{\xi_j}, \frac{1}{m_j})_j, D\right]$ (Definition 13.4) with $\xi_1 = \xi$. The basis $(e_n)_{n\in\mathbb{N}}$ of X dominates the basis $(y_n)_{n\in\mathbb{N}}$ of Y. Indeed, $G \subset D$ and the operator $Q : X \longrightarrow Y$ defined by $Q(e_n) = y_n$ satisfies $Q^*(y_n^*) = e_n^*$ and hence $Q^*(\overline{G_0}^{w^*}) = \overline{G}^{w^*} \subset B_{X^*}$, therefore $\|Q\| = 1$. It follows from Theorem 13.30 that the space X satisfies (ii),(iii),(iv).

Suppose now that Y^* is separable. Then the w^* closure of G_0 is norm separable in Y^* hence the w^* closure of G is norm separable in X^*, since $Q^*(\overline{G_0}^{w^*}) = \overline{G}^{w^*} \subset B_{X^*}$. Theorem 13.30 yields that $X^* = \overline{\text{span}}(\overline{G}^{w^*})$ therefore the dual space X^* is separable.

Suppose that Y is reflexive. Then $Y^* = \overline{\text{span}}\{y_n^* : n \in \mathbb{N}\}$ thus, since $Q^*(\overline{\text{span}}\{y_n^* : n \in \mathbb{N}\}) \subset \overline{\text{span}}\{e_n^* : n \in \mathbb{N}\}$ we get that $Q^*(Y^*) \subset X_*$. It follows that $\overline{G}^{w^*} \subset X_*$ hence $X^* \subset X_*$ therefore the space X is reflexive. $\square$

CHAPTER 14

Quotients of H.I. spaces

The results of this chapter concern the quotients of H.I. Banach spaces. We extend the results proved for ℓ^p spaces presented in chapter 11. What we consider worthwhile to be noticed is the precise description of the quotient map that could be useful in applications.

Fix a partition $(\Lambda_i)_{i\in\mathbb{N}}$ of $\mathbb{N}$, $\Lambda_i = \{\lambda_1^i < \lambda_2^i < \cdots\}$ and let $\mathcal{L}$ be the tree defined by this partition as in Definition 11.5.

DEFINITION 14.1. Let Z be a Banach space with a normalized bimonotone Schauder basis $(e_i)_{i\in\mathbb{N}}$. We set

$$G_Z^{\mathcal{L}} = \Big\{ E\sum_{i=1}^{d} a_i\Lambda_i^* \,:\, a_i\in\mathbb{Q} \text{ for } i=1,2,\dots d, \\ \sum_{i=1}^{d} a_i e_i^* \in \mathrm{B}_{Z^*},\ d\in\mathbb{N},\ E \text{ finite interval of } \mathbb{N} \Big\}.$$

(we remind that Eg denotes the restriction of the functional g to the interval E)

It is clear by the definition that the set $G_Z^{\mathcal{L}}$ is a ground subset of c_{00} (Definition 3.1). In the sequel we denote by Y the ground space $Y_{G_Z^{\mathcal{L}}}$ (Definition 3.2) by $(y_n)_{n\in\mathbb{N}}$ the natural Schauder basis of Y derived by the standard basis of c_{00}. It is easy to see that the closure of $G_Z^{\mathcal{L}}$ in the topology of pointwise convergence is the set

$$\overline{G_Z^{\mathcal{L}}}^p = \Big\{ E(w^* - \sum_{i=1}^{\infty} z^*(e_i)\Lambda_i^*) \,:\, z^*\in\mathrm{B}_{Z^*},\ \ E \text{ is an interval of } \mathbb{N} \Big\}.$$

We denote by E_m the interval of the natural numbers $E_m = \{m, m+1,\dots\}$.

PROPOSITION 14.2. Let Z be a Banach space with a normalized bimonotone Schauder basis $(e_n)_{n\in\mathbb{N}}$ and let Y be the ground space $Y_{G_Z^{\mathcal{L}}}$. If Z contains no isomorphic copy of ℓ^1 then also Y does not contain isomorphically ℓ^1.

Before passing to the proof of the proposition we prove the following lemma.

LEMMA 14.3. For every $m = 1,2,\dots$ and every choice of scalars $(a_i)_{i=1}^d$ we have that

$$\|\sum_{i=1}^{d} a_i E_m\Lambda_i^*\|_{Y^*} = \|\sum_{i=1}^{d} a_i e_i^*\|_{Z^*}.$$

Proof. Assume that $\|\sum_{i=1}^{d} a_i e_i^*\|_{Z^*} = 1$. Then the functional $f = \sum_{i=1}^{d} a_i E_m\Lambda_i^*$ belongs to the w^* closure of $G_Z^{\mathcal{L}}$ in Y^* hence $\|f\|_{Y^*} \le 1$.

We select scalars $(\tilde{a}_i)_{i=1}^d$ such that $\sum_{i=1}^d a_i\tilde{a}_i = (\sum_{i=1}^d a_i e_i^*)(\sum_{i=1}^d \tilde{a}_i e_i) = 1$ and $\|\sum_{i=1}^d \tilde{a}_i e_i\| = 1$. We choose inductively $l_i \in \Lambda_i$, $i = 1, 2, \ldots, d$ satisfying

$$m \le l_1 < l_2 < l_3 < \cdots < l_d$$

and we consider the vector $y \in Y$ defined as

$$y = \sum_{i=1}^d \tilde{a}_i y_{l_i}.$$

Then $f(y) = \sum_{i=1}^d a_i\tilde{a}_i = 1$ so to finish the proof it is enough to show that $\|y\| \le 1$ i.e. $g(y) \le 1$ for every $g \in G_Z^{\mathcal{L}}$.

Let $g \in G_Z^{\mathcal{L}}$; then g has the form $g = E\sum_{i=1}^{d'} \gamma_i \Lambda_i^*$ where E is an interval and $\|\sum_{i=1}^{d'} \gamma_i e_i^*\| \le 1$. By the definitions and the choice of $l_1, l_2, \ldots, l_d$ we have that if $E \cap \{l_1, l_2, \ldots, l_d\}$ is not empty then it is of the form $\{l_i : \ k_1 \le i \le k_2\}$ for some $1 \le k_1 \le k_2 \le \min\{d, d'\}$. By bimonotonicity we get that $\|\sum_{i=k_1}^{k_2} \tilde{a}_i e_i\| \le 1$ and $\|\sum_{i=k_1}^{k_2} \gamma_i e_i^*\| \le 1$ thus $\sum_{i=k_1}^{k_2} \gamma_i\tilde{a}_i = (\sum_{i=k_1}^{k_2} \gamma_i e_i^*)(\sum_{i=k_1}^{k_2} \tilde{a}_i e_i) \le 1$. Therefore $g(y) = \sum_{i=k_1}^{k_2} \gamma_i\tilde{a}_i \le 1$.

The proof is complete. □

Proof of Proposition 14.2. By the previous lemma, for each $m = 1, 2, \ldots$ the operator $T_m : Z^* \longrightarrow Y^*$ defined by $T_m(z^*) = w^* - \sum_{i=1}^{\infty} z^*(e_i) E_m \Lambda_i^*$ is an isometric embedding. The restriction of T_m to the unit ball of Z^* is $w^* - w^*$ continuous, as the w^* topology on bounded sets of Z^* (resp. Y^*) is determined by the pointwise convergence on the basis $(e_i)_{i\in\mathbb{N}}$ of Z (resp. the basis $(y_n)_{n\in\mathbb{N}}$ of Y). Thus by standard arguments (Banach Dieudonne theorem, see e.g. [**29**],[**30**]) we obtain that T_m is $w^* - w^*$ continuous, hence it is a dual operator. Denote by $Q_m : Y \longrightarrow Z$ the operator for which $Q_m^* = T_m$. Note that $Q_m(y_n) = 0$ if $n < m$ and $Q_m(y_n) = e_i$ if $n \ge m$ and $n \in \Lambda_i$.

To prove that ℓ^1 is not contained isomorphically in Y it is enough to find, for a given bounded sequence in Y, a weakly Cauchy subsequence. Let $(x_k)_{k\in\mathbb{N}}$ be a bounded sequence in Y. By Rainwater's Theorem [**45**] we have to find a subsequence $(x'_k)_{k\in\mathbb{N}}$ of $(x_k)_{k\in\mathbb{N}}$ such that $(y^*(x'_k))_{k\in\mathbb{N}}$ converges for every $y^* \in \mathrm{Ext}(\mathrm{B}_{Y^*})$. Since $\mathrm{B}_{Y^*} = \overline{\mathrm{conv}(G_Z^{\mathcal{L}})}^{w^*}$ we have that $\mathrm{Ext}(\mathrm{B}_{Y^*}) \subset \overline{G_Z^{\mathcal{L}}}^{w^*}$, therefore we have only to concern about the convergence of a subsequence for any $g \in \overline{G_Z^{\mathcal{L}}}^{w^*}$.

Since $\{(Q_m(x_k))_{k\in\mathbb{N}} \ \ m \in \mathbb{N}\}$ is a countable family of bounded sequences in Z, and Z contains no isomorphic copy of ℓ^1, by Rosenthal's ℓ^1 theorem and a diagonalization argument we may pass to a subsequence $(x'_k)_{k\in\mathbb{N}}$ of $(x_k)_{k\in\mathbb{N}}$ such that $(Q_m(x'_k))_{k\in\mathbb{N}}$ is a weakly Cauchy sequence in Z for each $m = 1, 2, \ldots$. Passing to a further subsequence we may assume that $(y_n^*(x'_k))_{k\in\mathbb{N}}$ is a convergent sequence for each $n = 1, 2, \ldots$ (recall that y_n^* are the biorthogonals of the basis $(y_n)_{n\in\mathbb{N}}$). We claim that $(x'_k)_{k\in\mathbb{N}}$ is the desired subsequence.

Indeed, let $g \in \overline{G_Z^{\mathcal{L}}}^{w^*}$. Then g is of the form $g = E(w^* - \sum_{i=1}^{\infty} z^*(e_i)\Lambda_i^*)$ where $z^* \in B_{Z^*}$ and E is an interval of $\mathbb{N}$. If E is a finite interval then g is the finite linear combination of some y_n^* and thus $(g(x'_k))_{k\in\mathbb{N}}$ is convergent. On the other hand if E is an infinite interval then $E = E_m$ for some m and it follows that $g = T_m z^*$. Therefore the sequence $(g(x'_k))_{k\in\mathbb{N}}$ is equal to $(z^*(Q_m x'_k))_{k\in\mathbb{N}}$ The later sequence converges since $(Q_m(x'_k))_{k\in\mathbb{N}}$ is a weakly Cauchy sequence in Z.

The proof of the proposition is complete. □

NOTATION 14.4. For the partition $(\Lambda_i)_{i\in\mathbb{N}}$ of $\mathbb{N}$ defined below we consider the linear map $Q : c_{00}(\mathbb{N}) \longrightarrow c_{00}(\mathbb{N})$ that extends the corresponding $e_n \mapsto e_i$ for all $n \in \Lambda_i$.

THEOREM 14.5. *Let Z be a Banach space containing no isomorphic copy of ℓ^1 and having a normalized bimonotone Schauder basis $(e_i)_{i\in\mathbb{N}}$. Then there exists an asymptotic ℓ^1 H.I. Banach space X with a normalized Schauder basis $(u_n)_{n\in\mathbb{N}}$ such that*

(i) *The above defined linear map $Q : (c_{00}(\mathbb{N}), \| \|_X) \longrightarrow (c_{00}(\mathbb{N}), \| \|_Z)$ is extended to a quotient map $Q : X \longrightarrow Z$.*
(ii) *The dual map $Q^* : Z^* \longrightarrow X^*$ is an isometric embedding and X^* is isomorphic to $X_* \oplus Z^*$ (here as before $X_* = \overline{\text{span}}\{u_n^* : n \in \mathbb{N}\}$ is the predual of X) with X_* H.I.*

Moreover the space X has the following properties:

(iii) *Every bounded linear operator R of X is of the form $R = \lambda I + W$ for some $\lambda \in \mathbb{R}$ and an operator W which is strictly singular and weakly compact.*
(iv) *The space X is somewhat reflexive.*

Proof. We begin with the definition of the space X. We denote by G the ground set $G_Z^{\mathcal{L}}$ as it is described in Definition 14.1 and Y_G denotes the ground space defined by the set G. Lemma 14.2 yields that ℓ^1 does not embed into Y_G. It follows from Corollary 13.36 that there exist a countable ordinal ξ such that G is an $\mathcal{S}_\xi$ bounded set. The space $X_G = T\left[G, (\mathcal{S}_{\xi_j}, \frac{1}{m_j})_j, D\right]$ with $\xi_1 = \xi$ is the desired space X (see Definition 13.4).

Indeed, Theorem 13.30 yields that the space X is an asymptotic ℓ^1 H.I. space hence its Schauder basis $(u_n)_{n\in\mathbb{N}}$ is boundedly complete and its predual X_* is also H.I. Furthermore assertions (iii) and (iv) in the statement of our theorem follow also by Theorem 13.30. It remains to prove (i) and (ii). For this we will use the next Lemma, which is analogous to Lemma 11.3.

LEMMA 14.6. For every $m = 1, 2, \ldots$ and every choice of scalars $(a_i)_{i=1}^d$ we have that

$$\|\sum_{i=1}^{d} a_i E_m \Lambda_i^*\|_{X^*} = \|\sum_{i=1}^{d} a_i e_i^*\|_{Z^*}$$

where E_m denotes the interval $E_m = \{m, m+1, \ldots\}$.

Proof. Assume that $\|\sum_{i=1}^{d} a_i e_i^*\|_{Z^*} = 1$. Then the functional $f = \sum_{i=1}^{d} a_i E_m \Lambda_i^*$ belongs to the w^* closure of the set $G = G_Z^{\mathcal{L}}$ in X^*, hence $f \in \overline{G}^{w^*} \subset \overline{D}^{w^*}$ and thus $\|f\|_{X^*} \leq 1$.

We select scalars $(\tilde{a}_i)_{i=1}^d$ such that $\sum_{i=1}^{d} a_i \tilde{a}_i = (\sum_{i=1}^{d} a_i e_i^*)(\sum_{i=1}^{d} \tilde{a}_i e_i) = 1$ and $\|\sum_{i=1}^{d} \tilde{a}_i e_i\| = 1$.

Let $\varepsilon > 0$; we'll show that $\|f\|_{X^*} > \frac{1}{1+\varepsilon}$. Choose $j \in \mathbb{N}$ such that

$$\frac{1}{m_j} < \frac{\varepsilon}{2d}.$$

We select $L^1, L^2, \ldots, L^d$ infinite subsets of $\mathbb{N}$, $L^i = (l_t^i)_{t\in\mathbb{N}}$ $i = 1, 2, \ldots, d$ satisfying the following conditions:

(i) $l_1^1 \geq m$.
(ii) $L^i \subset \Lambda_i$ for $i = 1, 2, \ldots d$.
(iii) $l_t^1 < l_t^2 < \cdots < l_t^d < l_{t+1}^1 < \cdots$ for each $t = 1, 2, \ldots$

Let $x_1 = \sum_{t=1}^{q} r_t u_{l_t^1} \in X$ be an $(\frac{\varepsilon}{2d}, j)$ B.S.C.C. Replacing the sets $L^1, L^2, \ldots, L^d$ by appropriate subsets we may assume that $\operatorname{supp} x_1$ is an initial segment of L^1. Set

$$x_i = \sum_{t=1}^{q} r_t u_{l_t^i} \quad \text{for } i = 1, 2, \ldots d.$$

and consider the vector

$$x = \sum_{i=1}^{d} \tilde{a}_i x_i$$

Since $E_m \Lambda_i^*(x_i) = 1$ for $i = 1, 2, \ldots, d$ we get that $f(x) = \sum_{i=1}^{d} a_i \tilde{a}_i = 1$ so it suffices to show that $\|x\| < 1 + \varepsilon$.

Claim. If $g \in G$ and $t \in \{1, 2, \ldots, q\}$ then $|g(\sum_{i=1}^{d} \tilde{a}_i u_{l_t^i})| \leq 1$.

Proof of the Claim. Let $g \in G$; then g is of the form $g = E \sum_{i=1}^{d'} \gamma_i \Lambda_i^*$ where E is an interval and $\|\sum_{i=1}^{d'} \gamma_i e_i^*\| \leq 1$. If the set $E \cap \{l_t^1, l_t^2, \ldots, l_t^d\}$ is not empty then it is of the form $\{l_t^i : k_1 \leq i \leq k_2\}$ for some $1 \leq k_1 \leq k_2 \leq \min\{d, d'\}$. By bimonotonicity we get that $\|\sum_{i=k_1}^{k_2} \tilde{a}_i e_i\| \leq 1$ and $\|\sum_{i=k_1}^{k_2} \gamma_i e_i^*\| \leq 1$ thus $|\sum_{i=k_1}^{k_2} \gamma_i \tilde{a}_i| = |(\sum_{i=k_1}^{k_2} \gamma_i e_i^*)(\sum_{i=k_1}^{k_2} \tilde{a}_i e_i)| \leq 1$. Therefore $|g(\sum_{i=1}^{d} \tilde{a}_i u_{l_t^i})| = |\sum_{i=k_1}^{k_2} \gamma_i \tilde{a}_i| \leq 1$ and this completes the proof of the claim. □

Since the norming set D of the space $X = T\left[G, \left(\mathcal{S}_{\xi_j}, \frac{1}{m_j}\right)_j, D\right]$ is a subset of the set $W = W[G, \left(\mathcal{S}_{\xi_j}, \frac{1}{m_j}\right)_j]$ the proof will be finished after showing that $f(\sum_{i=1}^{d} \tilde{a}_i x_i) \leq 1 + \varepsilon$ for every $f \in A$ (where $A \subset W$ is the set functionals having weight). Let $f \in A$. Using similar arguments to those in Definition 3.14 and Lemma 3.15 we may assume that there exists a tree $T_f = (f_a)_{a \in \mathcal{A}}$ of f (Definition 3.5) such that each f_a is not of type II (Definition 3.6). Let $(g_{a_s})_{s=1}^{s_0}$ be the functionals corresponding to the maximal elements of the tree $\mathcal{A}$. We denote by $\preceq$ the ordering of the tree $\mathcal{A}$. Let

$$A = \left\{ s \in \{1, 2, \ldots, s_0\} : \prod_{\gamma \prec a_s} \frac{1}{w(f_\gamma)} \leq \frac{1}{m_j} \right\}$$

$$B = \{1, 2, \ldots, s_0\} \setminus A$$

and set $f_A = f|_{\bigcup_{s \in A} \operatorname{supp} g_{a_s}}$, $f_B = f|_{\bigcup_{s \in B} \operatorname{supp} g_{a_s}}$.

We have that $|f_A(x_i)| \leq \frac{1}{m_j}$ for each i thus

$$|f_A(\sum_{i=1}^{d} \tilde{a}_i x_i)| \leq \frac{1}{m_j} \sum_{i=1}^{d} |\tilde{a}_i| \leq \frac{1}{m_j} \cdot d < \frac{\varepsilon}{2} \tag{40}$$

Since $x = \sum_{i=1}^{d} \tilde{a}_i x_i = \sum_{i=1}^{d} \tilde{a}_i (\sum_{t=1}^{q} r_t u_{l_t^i})$ the vector x takes also the form

$$x = \sum_{t=1}^{q} r_t (\sum_{i=1}^{d} \tilde{a}_i u_{l_t^i}).$$

Set

$$\begin{aligned} E_1 &= \Big\{ t \in \{1, 2, \ldots, q\} : \text{ the set } \{l_t^1, l_t^2, \ldots, l_t^d\} \text{ is contained in } \operatorname{ran} g_{a_s} \\ &\qquad \text{for some } s \in B \text{ or does not intersect any } \operatorname{ran} g_{a_s},\ s \in B \Big\} \\ E_2 &= \{1, 2, \ldots, q\} \setminus E_1. \end{aligned}$$

For each $s = 1, 2, \ldots, s_0$ set $\theta_s = \sum \left\{ r_t : \ \{l_t^1, l_t^2, \ldots, l_t^d\} \subset \operatorname{ran} g_{a_s} \right\}$ and observe that $\sum_{s \in B} \theta_s \leq 1$. By the Claim we have that

$$|g_{a_s}(\sum_{t \in E_1} r_t (\sum_{i=1}^{d} \tilde{a}_i u_{l_t^i}))| \leq \sum_{t \in E_1} r_t |g_{a_s}(\sum_{i=1}^{d} \tilde{a}_i u_{l_t^i})| \leq \sum_{t:\ \{l_t^1, l_t^2, \ldots, l_t^d\} \subset \operatorname{ran} g_{a_s}} r_t = \theta_s$$

therefore

$$|f_B(\sum_{t \in E_1} r_t (\sum_{i=1}^{d} \tilde{a}_i u_{l_t^i}))| \leq \sum_{s \in B} |g_{a_s}(\sum_{t \in E_1} r_t (\sum_{i=1}^{d} \tilde{a}_i u_{l_t^i}))| \leq \sum_{s \in B} \theta_s \leq 1. \tag{41}$$

By the definition of the set B we can prove, as in the proof of Lemma 3.16, that the family of functionals $\{g_{a_s} : s \in B\}$ is $\mathcal{S}_{\xi_{j-1}}^{q_j - 1}$ admissible (where $q_j = \log_2(m_j)$). Also, by the definition of the set E_2, for all $t \in E_2$ the set $\{l_t^1, l_t^2, \ldots, l_t^d\}$ intersects at least one ran g_{a_s}, $s \in B$ but is not contained in any of them. This implies that $\{l_t^1 : t \in E_2 \setminus \{\min E_2\}\} \in \mathcal{S}_{\xi_{j-1}}^{q_j-1}$ hence $\{l_t^1 : t \in E_2\} \in \mathcal{S}_{\xi_{j-1}}^{q_j} \subset (\mathcal{S}_{\xi_{j-1}+1}^M)^{q_j}$. The fact that $\sum_{t=1}^{q} r_t u_{l_t^1}$ is an $(\frac{\varepsilon}{2d}, j)$ B.S.C.C. implies that $\sum_{t\in E_2} r_t < \frac{\varepsilon}{2d}$. Therefore

$$f_B(\sum_{t\in E_2} r_t(\sum_{i=1}^{d} \tilde{a}_i u_{l_t^i})) \leq (\sum_{t\in E_2} r_t)(\sum_{i=1}^{d} |\tilde{a}_i|) < \frac{\varepsilon}{2d} \cdot d = \frac{\varepsilon}{2}. \tag{42}$$

From (40),(41) and (42) we conclude that

$$\begin{aligned} |f(x)| &\leq |f_A(\sum_{i=1}^{d} \tilde{a}_i x_i)| + |f_B(\sum_{t\in E_1} r_t(\sum_{i=1}^{d} \tilde{a}_i u_{l_t^i}))| + |f_B(\sum_{t\in E_2} r_t(\sum_{i=1}^{d} \tilde{a}_i u_{l_t^i}))| \\ &\leq \frac{\varepsilon}{2} + 1 + \frac{\varepsilon}{2} = 1 + \varepsilon. \end{aligned}$$

The proof of the lemma is complete. □

Completion of the proof of Theorem 14.5

The previous lemma immediately yields that Z^* isometrically embeds into X^*. Indeed, for $m = 1, 2, \ldots$ the operator $T_m : Z^* \longrightarrow X^*$ defined by the rule $T_m(z^*) = w^* - \sum_{i=1}^{\infty} z^*(e_i) E_m \Lambda_i^*$ is an isometric embedding. We denote by T the operator T_1. We also observe that T is $w^* - w^*$ continuous (this follows with arguments similar to the proof of Proposition 14.2), hence it is a dual operator. Let $Q : X \longrightarrow Z$ be such that $Q^* = T$. Observe that Q maps X onto Z and moreover $Q(u_n) = e_i$ for $n \in \Lambda_i$. This completes the proof of assertion (i) in the statement of the theorem.

Next we show that the quotient space X^*/X_* is isometric to Z^*. We consider the quotient map $R : X^* \longrightarrow X^*/X_*$ and we will show that the operator $R \circ T : Z^* \longrightarrow X^*/X_*$ is an onto isometry. Indeed, for $z^* \in Z^*$ we have that

$$\begin{aligned} \|R \circ T(z^*)\| &= \operatorname{dist}\left(w^* - \sum_{i=1}^{\infty} z^*(e_i)\Lambda_i^*, X_*\right) \\ &= \lim_m \operatorname{dist}\left(w^* - \sum_{i=1}^{\infty} z^*(e_i)\Lambda_i^*, \operatorname{span}\{u_1^*, u_2^*, \ldots, u_{m-1}^*\}\right) \\ &= \lim_m \|w^* - \sum_{i=1}^{\infty} z^*(e_i) E_m \Lambda_i^*\| = \lim_m \|T_m z^*\| \\ &= \|z^*\|. \end{aligned}$$

Thus $R \circ T$ is an isometry. Since $X^* = \overline{\operatorname{span}}(\overline{G}^{w^*})$ to see that $R \circ T$ is onto X^*/X_* it is enough to show that for every $g \in \overline{G}^{w^*}$ there exists $f \in Z^*$ such that $R \circ T(f) = g + X_*$. Such a g is of the form $g = E(w^* - \sum_{i=1}^{\infty} z^*(e_i)\Lambda_i)$ for some interval E and a $z^* \in B_{Z^*}$. If E is an infinite interval then the desired functional f coincides with z^*, while if E is a finite interval we take $f = 0$. Hence $R \circ T : Z^* \longrightarrow X^*/X^*$ is an isometry onto.

We conclude that Z^* is isometric to a complemented subspace of X^* and its complement is clearly the predual X_* of X. Therefore X^* is isomorphic to $Z^* \oplus X_*$.

The proof of the theorem is complete. □

A slight modification of the above construction yields the existence of quasi reflexive Hereditarily Indecomposable Banach spaces. Actually the following holds.

THEOREM 14.7. *Let Z be a finite dimensional Banach space with a bimonotone basis. Then there exists an H.I. Banach space Y with the quotient space Y^{**}/Y isometric to Z. In particular for each $k \in \mathbb{N}$ there exists a quasi reflexive and H.I. space Y with* $\dim Y^{**}/Y = k$.

Proof. Let $\dim Z = k$, and let $\{z_1, \dots, z_k\}$ be a normalized bimonotone basis of Z. We consider a partition of the set of natural numbers into k infinite disjoint sets $\Lambda_1, \dots, \Lambda_k$. We set

$$G = \{E(\sum_{i=1}^{k} a_i \Lambda_i) : a_1, \dots, a_k \in \mathbb{Q}, \sum_{i=1}^{k} a_i z_i \in \mathrm{B}_Z, \ E \text{ is a finite interval}\}.$$

It is easy to check that the set G is a ground subset of c_{00} and moreover that the ground space Y_G is isomorphic to c_0. In particular Y_G contains no isomorphic copy of ℓ^1 and satisfies the weak Banach-Saks property hence (Lemma 11.1) the ground set G is $\mathcal{S}_2$ bounded.

We consider the space $X_G = T[G, (\mathcal{S}_{n_j}, \frac{1}{m_j})_j, D]$. Theorem 9.8 yields that X_G is a H.I. space with a basis $(e_n)_{n \in \mathbb{N}}$ and its predual $(X_G)_* = \overline{\operatorname{span}}\{e_n^* : n \in \mathbb{N}\}$ is also H.I. The desired space Y is the space X_*. Indeed, following the proof of Theorem 14.5 we obtain that the quotient space $X_G^*/(X_G)_*$ is isometric to Z. $\square$

Since every H.I. space is not isomorphic to any of its proper subspaces the above theorem provides examples of quasi reflexive spaces Y which are not isomorphic to Y^{**}.

THEOREM 14.8. *Let Z be a separable Banach space that contains no isomorphic copy of ℓ^1. Then there exists a Banach space Y with a normalized bimonotone Schauder basis $(y_n)_{n \in \mathbb{N}}$ and a quotient map $Q : Y \longrightarrow Z$ such that ℓ^1 does not embed into Y, $\{Qy_n : n \in \mathbb{N}\}$ is a norm dense set of the unit sphere S_Z and Q^*Z^* is a complemented subspace of Y^*.*

Proof. Choose a bounded linear operator $R : \ell^1 \longrightarrow Z$ such that $\{R(e_n) : n \in \mathbb{N}\}$ is a norm dense subset of the unit sphere S_Z (here $(e_n)_{n \in \mathbb{N}}$ denotes the standard basis of ℓ^1). We set

$$W = \{Ef : f \in R^*(\mathrm{B}_{Z^*}), \ \ E \text{ interval of } \mathbb{N}\}.$$

and

$$\|x\|_W = \sup\{f(x) : f \in W\}$$

for every $x \in c_{00}$.

The following properties are easily established:

(i) For every $z^* \in Z^*$, $z^* \neq 0$ and every $\varepsilon > 0$ the set $\{n \in \mathbb{N} : R^*(z^*)(e_n) > \frac{\|z^*\|}{1+\varepsilon}\}$ is infinite.

(ii) The set W is a w^* compact subset of the unit ball of ℓ^∞ and moreover W is symmetric and closed under restrictions to intervals of $\mathbb{N}$.

(iii) Denoting by $(y_n)_{n \in \mathbb{N}}$ the standard basis of c_{00} we have that $\|y_n\|_W = 1$ and $\| \sum_{i=n}^{m} a_i y_i \|_W \leq \| \sum_{i=1}^{k} a_i y_i \|_W$ for all $m, n, k \in \mathbb{N}$ with $1 \leq n \leq m \leq k$ and every choice of scalars $(a_i)_{i=1}^k$.

The desired space Y is the completion of the space $(c_{00}(\mathbb{N}), \| \|_W)$. By (iii) $(y_n)_{n\in\mathbb{N}}$ is a normalized bimonotone Schauder basis of Y. Since for every $z^* \in B_{Z^*}$ we have $R^*(z^*) \in W \subset B_{Y^*}$ the operator $T : Z^* \longrightarrow Y^*$ defined by the rule $T(z^*) = R^*(z^*)$ is well defined and $\|T\| \le 1$. Property (i) yields that the operator T is an isometric embedding. Since also the restriction of T to the unit ball of Z^* is $w^* - w^*$ continuous, by standard arguments (Banach Dieudonne theorem) we get that T is $w^* - w^*$ continuous, hence it is a dual operator. We consider the operator $Q : Y \longrightarrow Z$ such that $T = Q^*$ and observe that Q maps Y onto Z. Since clearly $Qy_n = Re_n$ for all n, the set $\{Qy_n : n \in \mathbb{N}\}$ is norm dense in the unit sphere S_Z.

We next show that Y contains no isomorphic copy of ℓ^1. We consider a bounded sequence $(x_n)_{n\in\mathbb{N}}$ in Y and we will find a weakly Cauchy subsequence. Since $(Qx_n)_{n\in\mathbb{N}}$ is a bounded sequence in Z and Z contains no isomorphic copy of ℓ^1, by Rosenthal's ℓ^1 theorem, there exists $N \in [\mathbb{N}]$ such that the sequence $(Qx_n)_{n\in N}$ is a weakly Cauchy sequence in Z. By a standard diagonalization argument we may assume, passing to a further subsequence, that for all k the sequence $(y_k^*(x_n))_{n\in N}$ is convergent. We claim that the sequence $(x_n)_{n\in N}$ is weakly Cauchy. By Rainwater's theorem it is enough to show that the sequence $(y^*(x_n))_{n\in N}$ converges for every $y^* \in \operatorname{Ext} B_{Y^*}$. Since $B_{Y^*} = \overline{\operatorname{conv}(W)}^{w^*}$ we have that $\operatorname{Ext} B_{Y^*} \subset \overline{W}^{w^*}$. By the w^* compactness of the set W it suffices to prove that $(y^*(x_n))_{n\in N}$ is convergent for any $y^* \in W$. If such a functional y^* is finitely supported, the conclusion follows by the convergence of the sequences $(y_k^*(x_n))_{n\in N}$. If $y^* \in W$ is infinitely supported then $y^* = Q^*z^* - g$ for some $z^* \in B_{Z^*}$ and a finitely supported functional g, hence the conclusion follows from the fact that $(Qx_n)_{n\in N}$ is weakly Cauchy and the first case.

Set $Y_* = \overline{\operatorname{span}}\{y_n^* : n \in \mathbb{N}\}$ and consider the quotient map $L : Y^* \longrightarrow Y^*/Y_*$. We claim that the operator $L \circ Q^* : Z^* \longrightarrow Y^*/Y_*$ is an onto isometry. Indeed, by (i) and (iii) for $z^* \in Z^*$ we have that

$$\|L \circ Q^*(z^*)\| = \operatorname{dist}(Q^*(z^*), Y_*) = \lim_n \operatorname{dist}(Q^*z^*, \operatorname{span}\{y_1^*, y_2^*, \dots, y_{n-1}^*\}) = \|z^*\|.$$

Since the space Y contain no isomorphic copy of ℓ^1, by a theorem of Haydon [**32**], B_{Y^*} is the norm convex hull of its extreme points. The fact that $\operatorname{Ext} B_{Y^*} \subset W \subset B_{Y^*}$ yields that B_{Y^*} is the norm convex hull of W. Therefore, to see that $L \circ Q^*$ is onto we consider a $y^* \in W$ we find $f \in Z^*$ such that $L \circ Q^*(f) = y^* + Y_*$. If y^* is finitely supported we take $f = 0$. If y^* is infinitely supported, then it is of the form $y^* = Q^*z^* - g$ for some $z^* \in Z^*$ and a finitely supported functional g, and it is enough to take $f = z^*$. Hence, $L \circ Q^*$ is an isometry onto. We conclude that Z^* is isometric to Q^*Z^* and Y^* is isomorphic to $Y_* \oplus Z^*$.

The proof of the theorem is complete. □

Combining Theorems 14.5 and 14.8 we get the complete dichotomy for quotients of H.I. spaces.

THEOREM 14.9. *Let Z be a separable Banach space not containing ℓ^1. Then there exists a H.I. Banach space X and $Q : X \longrightarrow Z$ quotient map such that*

(i) *X has a Schauder basis $(x_n)_{n\in\mathbb{N}}$.*
(ii) *$X^* = W \oplus Q^*Z^*$.*
(iii) *If Z has a Schauder basis $(z_k)_{k\in\mathbb{N}}$ then Q maps $(x_n)_{n\in\mathbb{N}}$ onto $(z_k)_{k\in\mathbb{N}}$ and also the subspace W of X^* is H.I.*

(iv) *If no basis assumption is made for* Z *then* Q *could chosen such that* $\{Qx_n : n \in \mathbb{N}\}$ *be a norm dense subset of the unit sphere of* X.

As a corollary of Theorem 14.9 we obtain the following, which is related to results of [**19**].

COROLLARY 14.10. Let $(X_i)_{i=1}^n$ be a finite family of reflexive H.I. Banach spaces. Assume that each X_i has a Schauder basis. Then there exists a H.I. Banach space Y such that $Y^* = W \oplus (\oplus_{i=1}^n X_i)$ where W is also H.I. (and thus by Corollary 2 of [**19**] Y^* is HD_{n+1} according to the terminology inserted in [**19**]).

Proof. Since $(X_i)_{i=1}^n$ are reflexive spaces with Schauder bases the space $X = \oplus_{i=1}^n X_i^*$ has also a Schauder basis. By Theorem 14.9 there exists a H.I. space Y such that Y^* is isomorphic to $W \oplus X^* = W \oplus (\oplus_{i=1}^n X_i)$ where W is a H.I. subspace of Y^*. □

The following result concerns the weak approximation property (WAP), studied recently by E. Odell and H.-O. Tylli [**44**]. We recall from [**15**] that a Banach space X satisfies WAP if for every weakly compact subset W of X and every $\varepsilon > 0$ there exists a weakly compact operator $T \in \mathcal{L}(X)$ such that for every $x \in W$ we have that $\|x - Tx\| < \varepsilon$. The next proposition is a result of our communication with E. Odell who kindly permitted us to include it here.

PROPOSITION 14.11. Let G be an $\mathcal{S}_\xi$ bounded ground set such that the corresponding space X_G is nonreflexive. Then the space X_G fails WAP. In particular there exists a quasi reflexive X failing WAP.

Proof. As we have noticed in Remark 13.26 there exists a seminormalized weakly null sequence $(x_n)_{n\in\mathbb{N}}$ in X_G such that for every strictly singular $T \in \mathcal{L}(X_G)$ we have that $\|Tx_n\| \overset{n\to\infty}{\longrightarrow} 0$. Since X_G is nonreflexive and every $T \in \mathcal{L}(X_G)$ is of the form $T = \lambda I + S$ with S strictly singular we conclude that every weakly compact $T \in \mathcal{L}(X_G)$ is a strictly singular operator. Thus the sequence $(x_n)_{n\in\mathbb{N}}$ witnesses the failure of WAP in the space X_G. The required quasi reflexive space X failing WAP is any quasi reflexive space of the form X_G. □

REMARKS 14.12. 1. As we have mentioned before all the H.I. spaces $X_G = T[G, (\mathcal{S}_{\xi_j}, \frac{1}{m_j})_j, D]$ are asymptotic ℓ^1 spaces. It is an easy observation that the proofs concerning the quotient maps remain valid if the consider the spaces $Z_G = T[G, (\mathcal{S}_{\xi_j}, \frac{1}{m_j})_j]$. These spaces are also asymptotic ℓ^1, not containing ℓ^1 and unconditionally saturated. Therefore, every separable Banach space not containing an isomorphic copy of ℓ^1 is a quotient of an unconditionally saturated separable Banach space not containing ℓ^1.

2. The space $X_{G_0} = T[G_0, (\mathcal{S}_{n_j}, \frac{1}{m_j})_j, D]$ where $G_0 = \{\pm e_i : i \in \mathbb{N}\}$ is reflexive. As follows from Theorem 8.3 the dual space $X_{G_0}^*$ is also H.I. and it is easy to check that it is an asymptotic c_0 space. I. Deliyanni has recently shown that for all $1 < p < \infty$ there exist reflexive H.I. spaces which are asymptotic ℓ^p spaces [**18**].

3. Recently, and after submitting the present paper for publication, some new results related to H.I. spaces have occurred. The first concerns the existence of a reflexive indecomposable and unconditionally saturated Banach space X with few

operators (i.e. every $T : X \longrightarrow X$ is of the form $T = \lambda I + S$ with S strictly singular) [**12**]. Also in [**14**] a dual pair (X, X^*) of reflexive spaces has been provided with X unconditionally saturated and X^* H.I. A nonseparable reflexive Banach space not containing any unconditional basic sequence is given in [**11**].

Bibliography

[1] D. Alspach and S.A. Argyros, *Complexity of weakly null sequences*, Dissertationes Mathematicae, **321** (1992), 1–44.

[2] I. Amemiya and T. Ito, *Weakly null sequences in James spaces on trees*, Kodai Math. J., **4** (1981), 418–425.

[3] D. Amir and J. Lindenstrauss, *The structure of weakly compact sets in Banach spaces*, Ann. Math., **88** (1968), 35–46.

[4] G. Androulakis and E. Odell., *Distorting mixed Tsirelson spaces*, Israel J. Math., **109** (1999), 125–149.

[5] S.A. Argyros, *A universal property of reflexive Hereditarily Indecomposable Banach spaces*, Proc. AMS, **129** (2001), 3231–3239.

[6] S.A. Argyros and I. Deliyanni, *Examples of asymptotic ℓ^1 Banach spaces*, Transactions of the AMS, **349** (1997), 973–995.

[7] S.A. Argyros, I. Deliyanni, D. Kutzarova, and A. Manoussakis, *Modified mixed Tsirelson spaces*, J. of Func. Anal., **159** (1998), 43–109.

[8] S.A. Argyros, I. Deliyanni, and A. Manoussakis, *Distortion and spreading models in modified mixed Tsirelson spaces*, Studia Math., **157**(3) (2003), 199–236.

[9] S.A. Argyros and V. Felouzis, *Interpolating Hereditarily Indecomposable Banach spaces*, Journal of AMS, **13**(2) (2000), 243–294.

[10] S.A. Argyros and I. Gasparis, *Unconditional structures of weakly null sequences*, Transactions of AMS, **353** (2001), 2019–2058.

[11] S.A. Argyros, J. Lopez-Abad and S. Todorcevic, *A class of Banach spaces with no unconditional basic sequences*, C. R. Acad. Sci., Ser I, **337** (2003), 43–48.

[12] S.A. Argyros and A. Manoussakis, *An indecomposable and unconditionally saturated Banach space*, Studia Math., (to appear).

[13] S.A. Argyros, S. Mercourakis, and A. Tsarpalias, *Convex unconditionality and summability of weakly null sequences*, Isr. J. Math., **107** (1998), 157–193.

[14] S.A. Argyros and A. Tolias, *Indecomposability and unconditionality in duality*, Geom. and Funct. Analysis, (to appear).

[15] K. Astala and H.-O. Tylli., *Seminorms related to weak compactness and to Tauberian operators*, Math. Proc. Cambridge Phil. Soc., **107** (1990), 367–375.

[16] C. Bessaga and A. Pelczynski, *On bases and unconditional convergence of series in Banach spaces*, Studia Math., **17** (1958), 165–174.

[17] J. Bourgain, *On separable Banach spaces universal for all separable reflexive spaces*, Proc. of AMS, **79** (1980), 241–246. **79**, 241–246, 1980.

[18] I. Deliyanni, *Asymptotic ℓ^p H.I. spaces*, (in preparation).

[19] V. Ferenczi, *Hereditarily finitely decomposable Banach spaces*, Studia Math., **123**(2) (1997), 135–149.

[20] V. Ferenczi, *A uniformly convex Herediterily Indecomposable Banach space*, Israel J. Math., **102** (1997), 199–225.

[21] V. Ferenczi, *Quotient Hereditarily Indecomposable Banach spaces*, Canad. J. Math., **51** (1999), 566–584.

[22] T. Figiel and W.B. Johnson, *A uniformly convex Banach space which contains no ℓ^p*, Compositio Math., **29** (1974), 179–190.

[23] F. Galvin and K. Prikry, *Borel sets and Ramsey's theorem*, J. Symbolic Logic, **38** (1973), 193–198.

[24] I. Gasparis, *A dichotomy theorem for subsets of the powerset of the natural numbers*, Proc. AMS, **129** (2001), 759–764.

[25] I. Gasparis, *A continuum of totally incomparable Hereditarily Indecomposable Banach spaces*, Studia Math., **151** (2002), 277–298.

[26] W.T. Gowers, *A Banach space not containing c_0, ℓ^1 or a reflexive subspace*, Trans. Amer. Math. Soc., **344** (1994), 407–420.

[27] W.T. Gowers, *A new dichotomy for Banach spaces*, Geom. and Funct. Analysis, **6** (1996), 1083–1093.

[28] W.T. Gowers and B. Maurey, *The unconditional basic sequence problem*, Journal of AMS, **6** (1993), 851–874.

[29] P. Habala, P. Hajek, and V. Zizler, *Introduction to Banach Spaces*, vol. I, Matfyzpress, 1996.

[30] P. Habala, P. Hajek, and V. Zizler, *Introduction to Banach Spaces*, vol. II, Matfyzpress, 1996.

[31] J. Hagler, *A counterexample to several questions about Banach spaces*, Studia Math., **60** (1977), 289–308. **60**, 289–308, 1977.

[32] R.G. Haydon, *Some more characterizations of Banach spaces containing ℓ^1*, Math. Proc. Cambridge Phil. Soc., **80** (1976), 269–276.

[33] R.C. James, *A separable somewhat reflexive space with nonseparable dual*, Bull. Amer. Math. Soc., **80** (1974), 738–743.

[34] N. Kalton, *The basic sequence problem*, Studia Math., **116** (1995), 167–187.

[35] J. Lindenstrauss and C. Stegall, *Examples of separable spaces which do not contain ℓ^1 and whose duals are non-separable*, Studia Math., **54** (1975), 81–105.

[36] J. Lindenstrauss and L. Tzafriri, *Classical Banach spaces I, sequence spaces*, Springer Verlag 92, 1977.

[37] B. Maurey, *Banach spaces with few operators*, Handbook of the geometry of Banach spaces, vol. 2, edited by W.B. Johnson and J. Lindenstrauss, Elsevier, Amsterdam, 2002.

[38] S. Mazurkiewicz and W. Sierpinski, *Contribution a la topologie des ensembles denormables*, Fund. Math., **3** (1920), 17–27.

[39] S. Mercourakis, *On Cesaro summable sequences of continuous functions*, Mathematica, **42** (1995), 87–104.

[40] C. St. J. A. Nash-Williams, *On well quasi-ordering transfinite sequences*, Proc. Cambridge Phil. Soc., **61** (1965), 33–39.

[41] E. Odell, *A nonseparable Banach space not containing a subsymmetric basic sequence*, Isr. J. Math., **52** (1985), 97–109.

[42] E. Odell and H.P. Rosenthal, *A double-dual characterization of separable Banach spaces not containing ℓ^1*, Israel J. Math., **20** (1975), 375–384.

[43] E. Odell and T. Schlumprecht, *A Banach space block finitely universal for monotone bases*, Trans. Amer. Math. Soc., **352** (2000), 1859–1888.

[44] E. Odell and H.-O. Tylli, *Weak compact approximation in Banach spaces*, (in preparation).

[45] J. Rainwater, *Weak convergence of bounded sets*, Proc. AMS, **14** (1963), 999.

[46] H.P. Rosenthal, *A characterization of Banach spaces containing ℓ^1*, Proc. Nat. Acad. Sci., **71** (1974), 2411–2413.

[47] T. Schlumprecht, *An arbitrarily distortable Banach space*, Israel J. Math., **76** (1991), 81–95.

[48] S. Shelah, *A Banach space with few operators*, Israel J. Math., **30** (1978), 181–191.

[49] S. Shelah and J. Steprans, *A Banach space on which there are few operators*, Proc. AMS, **104** (1988), 101–105.

[50] B.S. Tsirelson, *Not every Banach space contains ℓ^p or c_0*, Funct. Anal. Appl., **8** (1974), 138–141.

[51] H.M. Wark, *A nonseparable reflexive Banach space on which there are few operators*, J. London Math. Soc., **64** (2001), 675–689.

Editorial Information

To be published in the *Memoirs*, a paper must be correct, new, nontrivial, and significant. Further, it must be well written and of interest to a substantial number of mathematicians. Piecemeal results, such as an inconclusive step toward an unproved major theorem or a minor variation on a known result, are in general not acceptable for publication. Papers appearing in *Memoirs* are generally longer than those appearing in *Transactions*, which shares the same editorial committee.

As of March 1, 2004, the backlog for this journal was approximately 4 volumes. This estimate is the result of dividing the number of manuscripts for this journal in the Providence office that have not yet gone to the printer on the above date by the average number of monographs per volume over the previous twelve months, reduced by the number of volumes published in four months (the time necessary for preparing a volume for the printer). (There are 6 volumes per year, each containing at least 4 numbers.)

A Consent to Publish and Copyright Agreement is required before a paper will be published in the *Memoirs*. After a paper is accepted for publication, the Providence office will send a Consent to Publish and Copyright Agreement to all authors of the paper. By submitting a paper to the *Memoirs*, authors certify that the results have not been submitted to nor are they under consideration for publication by another journal, conference proceedings, or similar publication.

Information for Authors

Memoirs are printed from camera copy fully prepared by the author. This means that the finished book will look exactly like the copy submitted.

The paper must contain a *descriptive title* and an *abstract* that summarizes the article in language suitable for workers in the general field (algebra, analysis, etc.). The *descriptive title* should be short, but informative; useless or vague phrases such as "some remarks about" or "concerning" should be avoided. The *abstract* should be at least one complete sentence, and at most 300 words. Included with the footnotes to the paper should be the 2000 *Mathematics Subject Classification* representing the primary and secondary subjects of the article. The classifications are accessible from `www.ams.org/msc/`. The list of classifications is also available in print starting with the 1999 annual index of *Mathematical Reviews*. The Mathematics Subject Classification footnote may be followed by a list of *key words and phrases* describing the subject matter of the article and taken from it. Journal abbreviations used in bibliographies are listed in the latest *Mathematical Reviews* annual index. The series abbreviations are also accessible from `www.ams.org/publications/`. To help in preparing and verifying references, the AMS offers MR Lookup, a Reference Tool for Linking, at `www.ams.org/mrlookup/`. When the manuscript is submitted, authors should supply the editor with electronic addresses if available. These will be printed after the postal address at the end of the article.

Electronically prepared manuscripts. The AMS encourages electronically prepared manuscripts, with a strong preference for AMS-LaTeX. To this end, the Society has prepared AMS-LaTeX author packages for each AMS publication. Author packages include instructions for preparing electronic manuscripts, the *AMS Author Handbook*, samples, and a style file that generates the particular design specifications of that publication series. Though AMS-LaTeX is the highly preferred format of TeX, author packages are also available in AMS-TeX.

Authors may retrieve an author package from e-MATH starting from `www.ams.org/tex/` or via FTP to `ftp.ams.org` (login as `anonymous`, enter username as password, and type `cd pub/author-info`). The *AMS Author Handbook* and the *Instruction Manual* are available in PDF format following the author packages link from `www.ams.org/tex/`. The author package can be obtained free of charge by sending email to `pub@ams.org` (Internet) or from the Publication Division, American Mathematical Society, 201 Charles St., Providence, RI 02904, USA. When requesting an author package, please specify $\mathcal{AMS}$-LaTeX or $\mathcal{AMS}$-TeX, Macintosh or IBM (3.5) format, and the publication in which your paper will appear. Please be sure to include your complete mailing address.

Sending electronic files. After acceptance, the source file(s) should be sent to the Providence office (this includes any TeX source file, any graphics files, and the DVI or PostScript file).

Before sending the source file, be sure you have proofread your paper carefully. The files you send must be the EXACT files used to generate the proof copy that was accepted for publication. For all publications, authors are required to send a printed copy of their paper, which exactly matches the copy approved for publication, along with any graphics that will appear in the paper.

TeX files may be submitted by email, FTP, or on diskette. The DVI file(s) and PostScript files should be submitted only by FTP or on diskette unless they are encoded properly to submit through email. (DVI files are binary and PostScript files tend to be very large.)

Electronically prepared manuscripts can be sent via email to `pub-submit@ams.org` (Internet). The subject line of the message should include the publication code to identify it as a Memoir. TeX source files, DVI files, and PostScript files can be transferred over the Internet by FTP to the Internet node `e-math.ams.org` (130.44.1.100).

Electronic graphics. Comprehensive instructions on preparing graphics are available at `www.ams.org/jourhtml/graphics.html`. A few of the major requirements are given here.

Submit files for graphics as EPS (Encapsulated PostScript) files. This includes graphics originated via a graphics application as well as scanned photographs or other computer-generated images. If this is not possible, TIFF files are acceptable as long as they can be opened in Adobe Photoshop or Illustrator. No matter what method was used to produce the graphic, it is necessary to provide a paper copy to the AMS.

Authors using graphics packages for the creation of electronic art should also avoid the use of any lines thinner than 0.5 points in width. Many graphics packages allow the user to specify a "hairline" for a very thin line. Hairlines often look acceptable when proofed on a typical laser printer. However, when produced on a high-resolution laser imagesetter, hairlines become nearly invisible and will be lost entirely in the final printing process.

Screens should be set to values between 15% and 85%. Screens which fall outside of this range are too light or too dark to print correctly. Variations of screens within a graphic should be no less than 10%.

Inquiries. Any inquiries concerning a paper that has been accepted for publication should be sent directly to the Electronic Prepress Department, American Mathematical Society, 201 Charles St., Providence, RI 02904, USA.

Editors

This journal is designed particularly for long research papers, normally at least 80 pages in length, and groups of cognate papers in pure and applied mathematics. Papers intended for publication in the *Memoirs* should be addressed to one of the following editors. In principle the Memoirs welcomes electronic submissions, and some of the editors, those whose names appear below with an asterisk (*), have indicated that they prefer them. However, editors reserve the right to request hard copies after papers have been submitted electronically. Authors are advised to make preliminary email inquiries to editors about whether they are likely to be able to handle submissions in a particular electronic form.

Titles in This Series

807 **Carlos A. Cabrelli, Christopher Heil, and Ursula M. Molter,** Self-similarity and multiwavelets in higher dimensions, 2004

806 **Spiros A. Argyros and Andreas Tolias,** Methods in the theory of hereditarily indecomposable Banach spaces, 2004

805 **Philip L. Bowers and Kenneth Stephenson,** Uniformizing dessins and Belyĭ maps via circle packing, 2004

804 **A. Yu. Ol'shanskii and M. V. Sapir,** The conjugacy problem and Higman embeddings, 2004

803 **Michael Field and Matthew Nicol,** Ergodic theory of equivariant diffeomorphisms: Markov partitions and stable ergodicity, 2004

802 **Martin W. Liebeck and Gary M. Seitz,** The maximal subgroups of positive dimension in exceptional algebraic groups, 2004

801 **Fabio Ancona and Andrea Marson,** Well-posedness for general 2×2 systems of conservation laws, 2004

800 **V. Poénaru and C. Tanasi,** Equivariant, almost-arborescent representations of open simply-connected 3-manifolds; A finiteness result, 2004

799 **Barry Mazur and Karl Rubin,** Kolyvagin systems, 2004

798 **Benoît Mselati,** Classification and probabilistic representation of the positive solutions of a semilinear elliptic equation, 2004

797 **Ola Bratteli, Palle E. T. Jorgensen, and Vasyl' Ostrovs'kyĭ,** Representation theory and numerical AF-invariants, 2004

796 **Marc A. Rieffel,** Gromov-Hausdorff distance for quantum metric spaces/Matrix algebras converge to the sphere for quantum Gromov-Hausdorff distance, 2004

795 **Adam Nyman,** Points on quantum projectivizations, 2004

794 **Kevin K. Ferland and L. Gaunce Lewis, Jr.,** The $RO(G)$-graded equivariant ordinary homology of G-cell complexes with even-dimensional cells for $G = \mathbb{Z}/p$, 2004

793 **Jindřich Zapletal,** Descriptive set theory and definable forcing, 2004

792 **Inmaculada Baldomá and Ernest Fontich,** Exponentially small splitting of invariant manifolds of parabolic points, 2004

791 **Eva A. Gallardo-Gutiérrez and Alfonso Montes-Rodríguez,** The role of the spectrum in the cyclic behavior of composition operators, 2004

790 **Thierry Lévy,** Yang-Mills measure on compact surfaces, 2003

789 **Helge Glöckner,** Positive definite functions on infinite-dimensional convex cones, 2003

788 **Robert Denk, Matthias Hieber, and Jan Prüss,** $\mathcal{R}$-boundedness, Fourier multipliers and problems of elliptic and parabolic type, 2003

787 **Michael Cwikel, Per G. Nilsson, and Gideon Schechtman,** Interpolation of weighted Banach lattices/A characterization of relatively decomposable Banach lattices, 2003

786 **Arnd Scheel,** Radially symmetric patterns of reaction-diffusion systems, 2003

785 **R. R. Bruner and J. P. C. Greenlees,** The connective K-theory of finite groups, 2003

784 **Desmond Sheiham,** Invariants of boundary link cobordism, 2003

783 **Ethan Akin, Mike Hurley, and Judy A. Kennedy,** Dynamics of topologically generic homeomorphisms, 2003

782 **Masaaki Furusawa and Joseph A. Shalika,** On central critical values of the degree four L-functions for GSp(4): The Fundamental Lemma, 2003

781 **Marcin Bownik,** Anisotropic Hardy spaces and wavelets, 2003

780 **S. Marmi and D. Sauzin,** Quasianalytic monogenic solutions of a cohomological equation, 2003

779 **Hansjörg Geiges,** h-principles and flexibility in geometry, 2003

For a complete list of titles in this series, visit the AMS Bookstore at **www.ams.org/bookstore/**.